PLASMA MODELLING: INNOVATIONS AND APPLICATIONS

Dr. Drs. Sugeng Rianto, M.Sc.
Prof. Dr. rer.nat. M. Nurhuda
Dr.rer.nat. Abdurrouf, S.Si.,M.Si.
Drs. Achmad Agus Dardiri, M.Si.
Drs. Didik Yudianto, M.Si.

PT. Frost Yunior
2024

PLASMA MODELLING: INNOVATIONS AND APPLICATIONS

Author :
Dr. Drs. Sugeng Rianto, M.Sc.
Prof. Dr. rer.nat. M. Nurhuda
Dr.rer.nat. Abdurrouf, S.Si.,M.Si.
Drs. Achmad Agus Dardiri, M.Si.
Drs. Didik Yudianto, M.Si.

ISBN : 978-623-97917-5-9

Cover Designer:
PT. Frost Yunior

Layout Designer:
PT. Frost Yunior

Prepress and Production:
PT. Frost Yunior

Publisher:
PT.Frost yunior

Editorial:
Jl.S.Parman gang merpati no 25 Pakis Krajan Banyuwangi

EXPERT'S FOREWORD

As a plasma expert, I am excited to provide insights into the fascinating field of plasma modeling. Plasma, often regarded as the fourth state of matter, is an ionized gas consisting of free electrons and ions that demonstrate unique behaviors, making it crucial in both natural and technological contexts. Understanding plasma dynamics is essential due to its widespread applications, ranging from energy generation in nuclear fusion to advanced materials processing in semiconductor manufacturing.

Plasma modeling allows scientists to predict the behavior of plasma under various conditions, aiding in innovation across industries. With the growth of computational power and numerical methods, plasma modeling has evolved, offering precise simulations of complex phenomena such as magnetohydro-dynamics (MHD), space weather forecasting, and fusion reactor design. These advancements enable the development of efficient plasma-based technologies in sectors like energy, medicine, and space exploration.

What makes plasma unique is its responsiveness to electromagnetic fields, creating a dynamic interplay between particles and fields. This behavior requires sophisticated mathematical models and computational tools, including particle-in-cell (PIC) methods and fluid models. By leveraging these approaches, plasma modeling has become an indispensable tool for scientists and engineers, pushing the boundaries of innovation and providing a clearer understanding of one of the most complex states of matter.

In conclusion, plasma modeling offers an exciting and challenging field that continues to evolve with technological advancements, promising breakthroughs in energy, space, and industrial applications.

Ir. D.J. Djoko Herry Santjojo M.Phil, Ph.D.

FOREWORD

The field of plasma modeling represents a fascinating intersection of physics, mathematics, and computational science, offering a crucial framework for understanding the behavior of plasma—the fourth state of matter. From applications in nuclear fusion to advancements in semiconductor technology, the role of plasma in modern scientific and industrial processes is undeniable.

This book, titled *Fundamentals and Applications of Plasma Modeling*, is designed to bridge the gap between the theoretical foundations of plasma physics and the practical aspects of computational modeling. By providing readers with a comprehensive exploration of both basic and advanced topics, it serves as an invaluable resource for students, researchers, and professionals.

Over the past few decades, the development of computational power and numerical methods has allowed for more precise simulations of plasma dynamics. The advancements detailed in this book aim to equip readers with the knowledge and tools necessary to navigate the challenges posed by plasma's complex, dynamic nature.

Whether you are a beginner or an experienced practitioner, this book offers a structured approach to mastering the art of plasma modeling, complete with case studies and practical examples. It is my hope that this resource inspires continued exploration and innovation in the field of plasma research.

Team Authors

LIST OF CONTENTS

LIST OF FIGURE

LIST OF TABLE

PLASMA

Lightening (Thunder)

"And the thunder exalts [Allah] with praise of Him – and the angels [as well] from fear of Him – and He sends thunderbolts and strikes with them whom He wills; yet they dispute about Allah, and He is severe in assault." **Al – Quran (13:13)**

(The behavior of lightning is an energetic and visually dramatic form of plasma, lightning ionizes the air it passes through, creating a channel of plasma)

Light Upon Light

"Allah is the Light of the heavens and the earth. The example of His light is like a niche within which is a lamp; the lamp is within glass, the glass as if it were a pearly [white] star lit from [the oil of] a blessed olive tree, neither of the east nor of the west, whose oil would almost glow even if untouched by fire. ***Light upon light****. Allah guides to His light whom He wills. And Allah presents examples for the people, and Allah is Knowing of all things."* **Al – Quran (24:35)**

(*Plasma* emits light through processes like electron transitions and thermal radiation. Plasma *is the source* of many natural and artificial light phenomena)

The classical interpretation of "***Light upon Light***" in the Qur'an (24:35) focuses on divine light, guidance, and spiritual insight. Interestingly, plasma—a highly energetic, light-emitting form of matter—shares some parallels with this concept of light as a dynamic, infinite force. Plasma is the source of much of the universe's visible light and energy, similar to how divine light is seen as the foundation of truth and spiritual understanding in Islam. By thinking of plasma in this way, we can reflect on the Qur'anic verse with a modern perspective, connecting the cosmic and spiritual aspects of light. This metaphor of plasma as "***Light upon Light***" helps us see the link between science and spirituality, showing how ancient wisdom can resonate with modern scientific ideas.

PREFACE: The Journey of Plasma Modeling

Plasma, often referred to as the fourth state of matter, holds a unique place in both the natural world and technological applications. From the stars that illuminate our night sky to the cutting-edge processes in semiconductor manufacturing, plasma is a versatile and ubiquitous presence. However, understanding and predicting the behavior of plasma presents significant challenges due to its complex and dynamic nature. This book, Fundamentals and Applications of Plasma Modelling, aims to bridge the gap between theoretical plasma physics and practical computational techniques, providing a comprehensive guide for students, researchers, and professionals in the field.

The journey of plasma modelling is as fascinating as it is intricate. The study of plasma involves a multidisciplinary approach, integrating principles of physics, mathematics, and computational science. Over the past few decades, advancements in computational power and numerical methods have revolutionized the field, enabling more accurate and detailed simulations of plasma behavior. This book is a culmination of these advancements, offering a detailed exposition of both the foundational concepts and the latest developments in plasma modelling.

One of the primary motivations for writing this book is the growing importance of plasma in various industrial and scientific applications. In industries ranging from electronics to aerospace, plasma processes are essential for innovation and efficiency. Similarly, in scientific research, understanding plasma phenomena is crucial for advancements in areas such as fusion energy, space exploration, and environmental science. By providing a structured and comprehensive overview of plasma modelling, this book seeks to empower readers with the knowledge and tools necessary to tackle these complex challenges.

The structure of the book is designed to cater to a wide audience, from beginners to advanced practitioners. It begins with an introduction to the basic concepts of plasma physics, ensuring that readers have a solid foundation before delving into more complex topics. Subsequent chapters cover the mathematical frameworks and numerical methods essential for plasma modelling, offering step-by-step guidance and practical insights. The inclusion of case studies and practical examples throughout the book helps to illustrate the real-world applications of the theoretical concepts discussed.

Another key aspect of this book is its focus on computational tools and software. In the modern era, the ability to effectively use simulation software is as important as understanding the underlying physics. This book provides an overview of popular plasma modelling software, along with practical tips on setting up simulations, analyzing results, and visualizing data. By combining theoretical knowledge with practical skills, readers will be well-equipped to conduct their own plasma simulations and contribute to advancements in the field.

The field of plasma modelling is continually evolving, driven by innovations in both science and technology. This book also explores emerging trends and future directions, including the integration of machine learning techniques and the development of quantum plasma models. By highlighting these cutting-edge developments, we aim to inspire readers to explore new frontiers and push the boundaries of what is possible in plasma research.

In conclusion, Fundamentals and Applications of Plasma Modelling is more than just a textbook; it is a comprehensive resource designed to guide and inspire the next generation of plasma scientists and engineers. Whether you are a student beginning your journey in plasma physics, a researcher seeking to deepen your understanding, or a professional looking to apply plasma modelling in your work, this book offers valuable insights and practical guidance. We hope that you find this book both informative and inspiring, and that it serves as a catalyst for your own contributions to the fascinating world of plasma.

Importance and Relevance of Plasma Modelling

Plasma modelling holds significant importance in both scientific research and industrial applications due to the unique properties and versatile nature of plasma. Plasma, often described as the fourth state of matter, consists of a collection of free-moving ions and electrons, making it highly responsive to electromagnetic fields. This responsiveness allows plasma to be used in various advanced technologies, from energy generation to materials processing.

In the field of energy, plasma modelling is critical for the development of nuclear fusion reactors, which promise a nearly limitless source of clean energy. Fusion processes, which power the sun and other stars, can be replicated on Earth using high-temperature plasmas. Accurate plasma models help scientists understand and control these processes, improving reactor designs and bringing us closer to achieving practical fusion energy.

Industrial applications of plasma include semiconductor manufacturing, where plasmas are used in processes like etching and deposition to create intricate microelectronic circuits. Plasma modelling enables precise control over these processes, leading to higher efficiency and better-quality products. In materials science, plasmas are employed to modify surfaces, deposit thin films, and enhance material properties. Understanding plasma interactions through modelling allows for optimization and innovation in these applications.

Environmental applications of plasma technology are also growing. Plasmas can break down pollutants in air and water, providing a means for pollution control and waste management. Plasma modelling aids in designing systems that maximize pollutant degradation while minimizing energy consumption, making these technologies more viable and cost-effective.

Moreover, plasma modelling is essential in space science and astrophysics. Space plasmas, such as the solar wind and planetary magnetospheres, affect satellite operations and space weather. Accurate models help predict and mitigate these effects, ensuring the safety and reliability of space missions.

In medicine, plasma technology is emerging as a powerful tool for sterilization, wound healing, and cancer treatment. Plasma modelling contributes to understanding the biological effects of plasmas, leading to safer and more effective medical devices and therapies.

Overall, plasma modelling is a cornerstone for advancing technology and scientific understanding across multiple domains. By providing insights into plasma behavior and interactions, it enables innovation, improves efficiency, and helps solve complex problems in energy, industry, environment, space, and health.

Brief History and Advancements in Plasma Modelling

The history of plasma modelling is a tale of scientific curiosity and technological progress. The study of plasma began in the late 19th and early 20th centuries with early observations of ionized gases. However, it wasn't until the mid-20th century that plasma physics emerged as a distinct field, driven by the development of nuclear fusion research and advancements in space science.

In the 1950s and 1960s, the quest for controlled nuclear fusion sparked significant interest in plasma modelling. Researchers developed basic theoretical models to understand plasma confinement and stability in fusion reactors. The advent of computers during this period revolutionized plasma research, allowing for more complex simulations and calculations. Early computational models focused on magnetohydrodynamics (MHD), describing plasma as a fluid influenced by magnetic fields.

The 1970s and 1980s saw further advancements in plasma modelling with the development of more sophisticated numerical methods. The introduction of the Particle-in-Cell (PIC) method was a major milestone, enabling simulations of plasma at the kinetic level. This approach allowed researchers to study phenomena such as plasma waves, instabilities, and particle interactions in greater detail.

During the 1990s, the field of plasma modelling expanded beyond fusion research to encompass a wider range of applications. Advances in computational power and algorithms facilitated the development of models for industrial processes, space plasmas, and environmental applications. Researchers began integrating different modelling approaches, combining fluid and kinetic descriptions to capture multi-scale plasma behavior.

In the 21st century, plasma modelling has continued to evolve with the advent of high-performance computing and sophisticated software. Modern simulations can handle complex geometries, multi-physics interactions, and large-scale systems with high accuracy. The use of parallel computing and advanced numerical techniques has pushed the boundaries of what is possible in plasma modelling.

Recent advancements include the incorporation of machine learning and data-driven approaches. These techniques are being used to enhance model accuracy, optimize simulations, and uncover new insights from large datasets. Additionally, the development of open-source plasma modelling software has democratized access to powerful tools, enabling broader participation in plasma research.

The history of plasma modelling is marked by continuous innovation and interdisciplinary collaboration. From its early theoretical foundations to the cutting-edge simulations of today, plasma modelling has played a crucial role in advancing our understanding of plasmas and harnessing their potential for various applications. As technology continues to advance, plasma modelling will remain a dynamic and essential field, driving progress in science and engineering.

Overview of Book Structure and Content

Fundamentals and Applications of Plasma Modelling is structured to provide a comprehensive guide to the principles, techniques, and applications of plasma modelling. The book is designed to cater to a diverse audience, including students, researchers, and professionals, offering both foundational knowledge and advanced insights.

Chapter 1: Introduction to Plasma Physics

The book begins with an introduction to the fundamental

concepts of plasma physics. This chapter covers the definition and characteristics of plasma, types of plasma, and key parameters such as temperature, density, and potential. It also explores the formation and sources of plasma, both natural and artificial, and discusses the various applications of plasma in industry, medicine, and the environment.

Chapter 2: Mathematical Foundations of Plasma Modelling

This chapter delves into the mathematical frameworks essential for plasma modelling. It introduces fundamental equations such as Maxwell's equations and fluid equations, and explains the kinetic theory, including the Boltzmann equation and distribution functions. Various plasma approximation methods, including magnetohydrodynamics (MHD), two-fluid, multi-fluid models, and particle-in-cell (PIC) methods, are thoroughly examined.

Chapter 3: Numerical Methods in Plasma Modelling

Numerical methods form the backbone of plasma modelling, and this chapter provides an in-depth look at discretization techniques, including finite difference, finite element, and finite volume methods. It discusses solving linear and nonlinear equations using iterative and direct solvers, and covers stability and convergence issues, such as time-stepping methods, the CFL condition, and error analysis.

Chapter 4: Computational Tools and Software for Plasma Modelling

This chapter offers a comprehensive overview of popular plasma modelling software, both commercial and open-source. It guides readers through the process of setting up a plasma simulation, defining the problem domain, and specifying initial and boundary conditions. Mesh generation and refinement techniques are also covered. The chapter concludes with post-processing and visualization techniques, helping readers analyze and present their simulation results effectively.

Chapter 5: Case Studies in Plasma Modelling

Practical applications are highlighted through detailed case studies in various fields. This chapter examines industrial plasma processes such as etching and deposition, plasma dynamics in space and astrophysics, and environmental and medical applications. Each case study provides real-world examples of how plasma modelling is applied to solve complex problems and innovate new solutions.

Chapter 6: Advanced Topics in Plasma Modelling

For readers seeking to explore more complex aspects of plasma modelling, this chapter covers advanced topics such as nonlinear plasma phenomena, including turbulence and instabilities, and multiscale modelling approaches. Emerging trends like machine learning in plasma modelling and quantum plasmas are also discussed, providing insights into the future direction of the field.

Chapter 7-12: Practical Applications and Future Trends

The final chapter focuses on the real-world applications of plasma modelling, highlighting success stories and practical insights from industry and research. It explores future trends and innovations in computational methods and interdisciplinary integration, and addresses current challenges and opportunities in the field.

Appendices

The book includes several appendices, providing a glossary of key terms, a mathematical reference with important formulas and derivations, and a guide to popular plasma modelling software with tutorials and resources for beginners.

References and Index

A comprehensive list of references and a detailed index are provided to facilitate further reading and quick navigation through the book. Overall, Fundamentals and Applications of Plasma Modelling offers a structured and detailed exploration of plasma modelling, combining theoretical foundations with practical applications to serve as a valuable resource for anyone involved in the field of plasma research and technology.

1: Introduction to Plasma Physics

1.1 Basic Concepts of Plasma

Plasma, often referred to as the fourth state of matter, is a distinct phase of matter characterized by its unique properties and behaviors. Unlike solids, liquids, and gases, plasma is composed of charged particles—ions and electrons—that are not bound together but move freely. This section delves into the fundamental concepts of plasma, exploring its definition, characteristics, types, and key parameters.

1.1.1. Definition and Characteristics

Plasma is defined as a quasi-neutral gas of charged and neutral particles that exhibits collective behavior [1]. The term "quasi-neutral" indicates that the overall charge density is nearly zero, with the number of positive charges (ions) approximately equal to the number of negative charges (electrons). This balance of charges imparts plasma with unique properties, such as its response to electromagnetic fields.

One of the defining characteristics of plasma is its ability to conduct electricity. Due to the presence of free-moving charged particles, plasmas are excellent conductors and can support electric currents. This property is exploited in various technological applications, from fluorescent lighting to plasma televisions [2].

Plasmas also exhibit collective behaviors, where the motions of individual particles are influenced by electric and magnetic fields generated by the particles themselves. These collective effects give rise to phenomena such as plasma oscillations, waves, and instabilities [3]. The study of these behaviors is crucial for understanding and controlling plasma in practical applications.

1.1.2. Types of Plasma

Plasmas can be broadly categorized into two main types: thermal (or equilibrium) plasmas and non-thermal (or non-equilibrium) plasmas.

ke

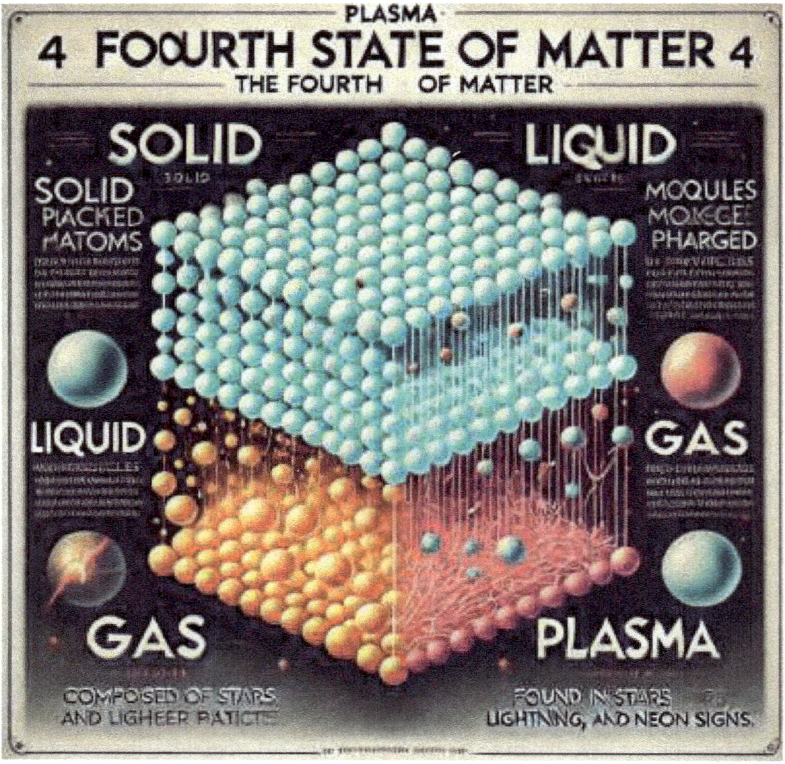

Figure 1. The schematic image illustrating plasma as the fourth state of matter.

Thermal Plasmas: In thermal plasmas, the temperatures of electrons, ions, and neutral particles are nearly equal, reaching equilibrium. These plasmas are typically very hot, with temperatures ranging from thousands to millions of degrees Kelvin. Examples of thermal plasmas include the sun and other stars, as well as artificial plasmas generated in welding torches and plasma arcs [4].

Non-Thermal Plasmas: In non-thermal plasmas, there is a

significant temperature difference between electrons and the heavier ions and neutral particles. Electrons are much hotter than the ions and neutrals, resulting in a non-equilibrium state. These plasmas can occur at relatively low temperatures and are often used in applications where high-energy electrons are needed without heating the entire gas. Examples include fluorescent lamps, plasma TVs, and certain types of plasma-based air purifiers [5].

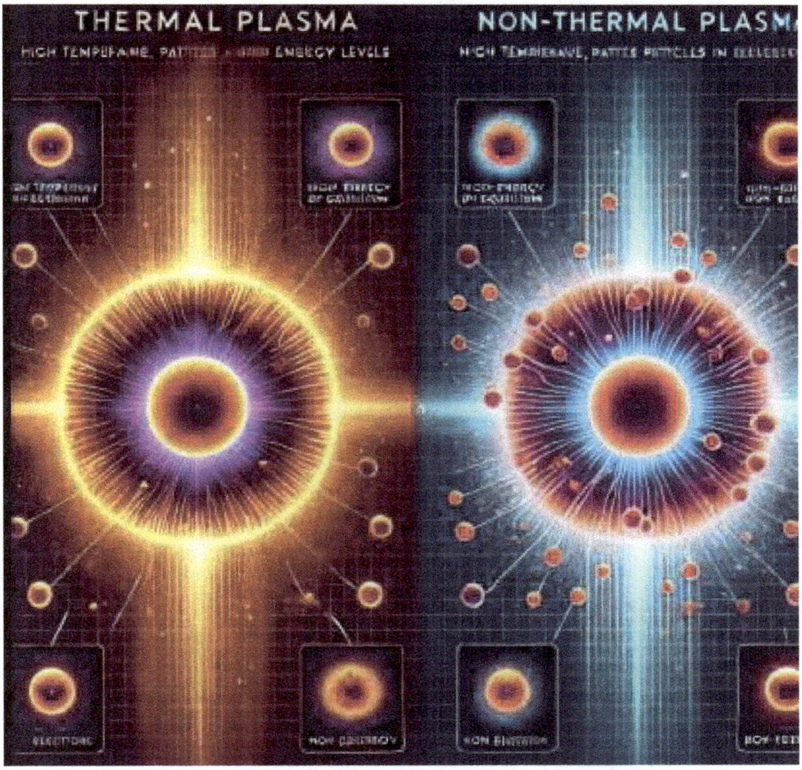

Figure 2. the schematic image depicting the two main types of plasmas: thermal (equilibrium) plasmas and non-thermal (non-equilibrium) plasmas

1.1.3. Plasma Parameters

Several key parameters characterize the behavior and properties of plasma. Understanding these parameters is essential for both theoretical studies and practical applications.

Temperature: Plasma temperature can refer to the kinetic energy of electrons, ions, or neutral particles. In many plasmas, electron temperature (T_e) is much higher than ion temperature (T_i). Temperature is typically measured in electron volts (eV) or Kelvin (K). High temperatures are required to ionize gases and maintain a plasma state, particularly in thermal plasmas [6].

Density: Plasma density refers to the number of particles per unit volume, usually measured in particles per cubic centimeter (cm^3). Electron density (n_e) and ion density (n_i) are crucial for determining plasma behavior and its interactions with electromagnetic fields. High-density plasmas are found in stars, while low-density plasmas exist in space and certain laboratory conditions [7].

Debye Length: The Debye length (λ_D) is a measure of the distance over which electric potentials are shielded out by the plasma. It is given by the formula:

$$\lambda_D = \sqrt{\frac{\epsilon_0 k_B T_e}{n_e e^2}}$$

where $\epsilon_0$is the permittivity of free space, k_B is the Boltzmann constant, $\boldsymbol{T_e}$ is the electron temperature, n_e is the electron density, and e_e is the elementary charge [8]. The Debye length indicates the scale over which the plasma can shield electric fields, playing a critical role in defining plasma behavior and its interaction with external fields.

Plasma Frequency: Plasma frequency (ω_p) is the natural oscillation frequency of electrons in the plasma. It is given by the formula:

$$\omega_p = \sqrt{\frac{n_e e^2}{m_e \epsilon_0}}$$

where m_e is the electron mass [9]. The plasma frequency determines the response of plasma to electromagnetic waves and is important for understanding wave propagation in plasmas.

Magnetic Field: Many plasmas exist in the presence of magnetic fields, which significantly influence their behavior. The strength and orientation of the magnetic field can affect plasma confinement, stability, and wave propagation. The interaction of plasmas with magnetic fields is a critical aspect of many applications, from magnetic confinement fusion to space plasmas [10].

1.1.4. Plasma Formation and Sources

Plasmas can be found in both natural and artificial environments. Natural plasmas include stellar interiors, the solar wind, and lightning. The sun, for example, is a massive ball of hot plasma, with nuclear fusion reactions occurring in its core, producing the energy that powers our solar system [11]. Lightning is a transient plasma phenomenon where a high-voltage electric discharge ionizes the air, creating a plasma channel that allows current to flow [12].

Artificial plasmas are generated in a variety of devices for numerous applications. In industrial settings, plasmas are used for processes such as plasma cutting, welding, and surface treatment. These plasmas are typically generated using electric arcs or radiofrequency (RF) discharges. In electronics, plasma etching is a crucial technique for fabricating microchips, where plasmas precisely remove material from semiconductor wafers [13].

In the medical field, non-thermal plasmas are employed for sterilization and tissue treatment. These plasmas can effectively kill bacteria and promote wound healing without damaging surrounding tissues [14]. Environmental applications include plasma-based systems for air and water purification, where plasmas break down pollutants into less harmful substances [15].

1.1.5. Applications of Plasma

The applications of plasma are vast and diverse, spanning numerous fields and industries. In addition to the aforementioned industrial, medical, and environmental uses, plasmas play a critical role in the field of fusion energy. Controlled thermonuclear fusion, which aims to replicate the energy-producing processes of the sun, relies on magnetic confinement and inertial confinement of high-temperature plasmas. These methods seek to achieve the conditions necessary for sustained fusion reactions, offering the potential for a nearly limitless and clean energy source [16].

In space exploration, understanding plasma behavior is essential for protecting spacecraft and satellites from the harsh plasma environment of space. The study of space plasmas, including the solar wind and Earth's magnetosphere, helps scientists predict space weather and its impacts on communication and navigation systems [17].

Furthermore, plasmas are used in the entertainment industry, such as in plasma displays for televisions and signage. These displays leverage the light-emitting properties of plasmas to produce high-quality images with vibrant colors and high contrast [18].

In summary, the basic concepts of plasma encompass its definition, characteristics, types, and key parameters, all of which are fundamental to understanding and harnessing plasma for various scientific and technological applications. From industrial processes to space exploration, plasmas offer immense potential and continue to be a focus of research and innovation.

1.2. Plasma Physics and Distribution Functions

Plasma physics involves the study of ionized gases composed of charged particles such as ions and electrons, where understanding particle distribution is crucial due to its deviation from classical equilibrium distributions, caused by unique interactions in plasma environments. These distributions, particularly in terms of velocity and energy, play a key role in explaining plasma behavior, influencing phenomena like stability and wave propagation. Several essential distribution functions, including the **Maxwell-Boltzmann**, **Druyvesteyn**, and **Kappa** distributions, are used to describe plasma systems, and their significance is explored across various applications in plasma physics.

Different distribution functions are essential in plasma physics to capture the diverse behaviors of particles under various conditions. Maxwell-Boltzmann is appropriate for equilibrium plasmas, while the Druyvesteyn and kappa distributions are more suited to non-equilibrium systems. Each function plays a crucial role in understanding and modeling plasma behavior across different applications.

In brieft understanding velocity distributions for a plasma can be done using:

- **Maxwell-Boltzmann distribution** (classical thermal distribution)
- **Druyvesteyn distribution** (non-equilibrium plasma)
- **Kappa distribution** (non-Maxwellian plasma for high-energy particles).

Comparing the velocity distributions for particles in a plasma using the **Maxwell-Boltzmann**, **Druyvesteyn**, and **Kappa** distributions can conclude that:

- The **Maxwell-Boltzmann distribution** shows the classical thermal spread of velocities, centered around zero with fewer particles at higher velocities.
- The **Druyvesteyn distribution** has a flatter shape, reflecting fewer high-velocity particles and a broader range of lower-energy particles.
- The **Kappa distribution** has a noticeable high-energy tail, which is typical in non-equilibrium plasmas where more particles have high velocities than predicted by Maxwellian distributions.

In plasma physics, understanding particle distribution functions is essential for describing the behavior of particles, including their energy, velocity, and density distributions. Different types of distribution functions provide insights into how particles interact and evolve under various plasma conditions. Below are some key distribution functions commonly used in plasma studies:

1.2.1. Maxwell-Boltzmann Distribution

The Maxwell-Boltzmann distribution is the most familiar and widely applied distribution in classical statistical mechanics. It assumes that the particles in the plasma are in thermal equilibrium and describes the velocity distribution of particles in such a scenario. This distribution is typically used in high-temperature plasmas, where particle collisions dominate and the system is near equilibrium.

The **Maxwell-Boltzmann distribution** is one of the most widely used models for describing particle velocity distributions in thermal equilibrium. In this distribution, particles in a gas (or plasma) spread their velocities in a way that most particles have speeds close to a mean value, with fewer particles having much higher or lower velocities. This distribution is appropriate when the plasma is in equilibrium and is used for analyzing various classical plasma behaviors. However, it may fail in situations involving strong electromagnetic fields or non-equilibrium states

- The Maxwell-Boltzmann distribution describes the probability $P(v)$ of a particle having a velocity v as:

  $P(v) \propto e^{-\frac{mv^2}{2k_B T}}$

  where m is the mass of the particle, v is the velocity, k_B is the Boltzmann constant, and T is the temperature [1].

While this distribution provides a good description for many plasma conditions, its accuracy diminishes in non-equilibrium or low-pressure plasmas where electron-electron collisions are less frequent[2]. In plasma systems like **dielectric barrier discharges**, assuming a Maxwellian distribution can lead to overestimated values for key parameters like ionization rates and electron mobility[3].

1.2.2. Druyvesteyn Distribution

The Druyvesteyn distribution function is used to describe electron velocity distributions in plasmas where non-equilibrium conditions dominate, particularly in low-pressure or low-temperature plasmas. It characterizes a situation where collisions are infrequent, and the distribution exhibits a flatter curve at low velocities compared to the Maxwell-Boltzmann distribution. This distribution is often applied in radio-frequency (RF) discharges or dielectric barrier discharges (DBDs).

The **Druyvesteyn distribution** is used to describe non-

Maxwellian plasmas, especially in **low-pressure plasmas**. In contrast to the Maxwell-Boltzmann distribution, the Druyvesteyn function results in fewer high-energy particles, reflecting a more flattened electron energy distribution. This is crucial when studying electromagnetic wave propagation in bounded plasmas and low-pressure discharge plasmas. The Druyvesteyn distribution significantly influences plasma properties like electron mobility and ion acoustic wave dispersion [4].

- The Druyvesteyn distribution function is mathematically flatter at low energies and more peaked at higher energies, making it more appropriate for plasmas that deviate from thermal equilibrium

  $$f(v) \propto v^2 e^{-\alpha v^4}$$

  where α is a constant related to the energy of the particles[4]

Studies have shown that the Druyvesteyn distribution is better suited for modeling non-equilibrium electron distributions, particularly in RF discharges where electrons do not maintain a Maxwellian distribution[3]. Moreover, the **electron density** in low-pressure plasmas is often governed by this distribution, providing better accuracy than the Maxwellian distribution in modeling plasmas' interaction with electromagnetic waves[5].

1.2.3. Bi-Maxwellian and Bi-Druyvesteyn Distributions

In some plasmas, especially space plasmas, the electron population can be divided into two distinct groups: a low-energy population and a high-energy "tail." To model these systems, researchers use bi-Maxwellian or bi-Druyvesteyn distributions, where two separate distributions describe the different electron populations.

- **Bi-Maxwellian Distribution**: Two Maxwellian distributions are used to describe a system where electrons at different energy levels have different temperatures.
- **Bi-Druyvesteyn Distribution**: This is similar to the bi-Maxwellian, but Druyvesteyn distributions are used to better describe the low-energy and high-energy electron populations

in non-equilibrium plasmas[3].

1.2.4. Kappa Distribution

The kappa distribution is often used in space plasmas where high-energy particles form a significant fraction of the overall particle population, causing a departure from the Maxwell-Boltzmann distribution. This distribution accounts for the presence of high-energy tails by introducing a parameter κ that governs the strength of the tail.

- The kappa distribution is defined as:

$$f(v) \propto \left(1 + \frac{v^2}{\kappa v_{th}^2}\right)^{-(\kappa+1)}$$

where v_{th} is the thermal velocity, and κ determines the tail's slope. Larger κ values correspond to distributions closer to Maxwellian [6].

The **Kappa distribution** is commonly used to describe non-equilibrium, steady-state plasmas, especially in space plasmas. This distribution is characterized by a high-energy tail, representing more high-energy particles than the Maxwellian model predicts. Kappa distributions are found in a wide variety of environments, including space and laboratory plasmas. They arise due to deviations from equilibrium and are crucial for describing particles influenced by long-range interactions or low-collision-rate environments. These distributions are particularly useful for describing plasmas in **fusion experiments** and are gaining popularity in the modeling of laboratory plasmas[6].

1.2.5. Other Distribution Functions

Apart from the above distributions, there are several other specialized distribution functions used in plasma physics:

- **Bi-Maxwellian Distribution**: This distribution describes systems where two distinct electron populations exist with different temperatures, often seen in systems like **inductively coupled plasmas**. It is used when both a cold and a hot electron population are present and need to be accounted for separately[2].

- **Maxwell-Boltzmann–Jüttner Distribution**: This distribution is an extension of the Maxwell-Boltzmann model for relativistic particles. It is applied to relativistic plasmas where particle velocities approach the speed of light, leading to deviations from the classical Maxwellian distribution[7].
- **Bi-Druyvesteyn Distribution**: Similar to the bi-Maxwellian distribution, this model accounts for systems with two different Druyvesteyn-like populations. It is particularly useful for modeling systems where both low and high-energy electron groups deviate from equilibrium [3].

Understanding plasma behavior requires accurate models of particle distributions. While the **Maxwell-Boltzmann distribution** is widely used for equilibrium systems, more complex distributions like the **Druyvesteyn** and **Kappa distributions** are necessary for non-equilibrium and low-pressure plasma environments. These distribution functions provide essential insights into plasma properties, influencing everything from wave propagation to plasma stability.

Different distribution functions are essential in plasma physics to capture the diverse behaviors of particles under various conditions. Maxwell-Boltzmann is appropriate for equilibrium plasmas, while the Druyvesteyn and kappa distributions are more suited to non-equilibrium systems. Each function plays a crucial role in understanding and modeling plasma behavior across different applications.

1.3. Plasma in Space

Plasma, often referred to as the "fourth state of matter," is the most common form of visible matter in the universe and behaves very differently in space compared to Earth due to the unique environmental conditions. Plasma in space, such as the solar wind or plasma in interstellar environments, is primarily governed by electromagnetic forces, which can lead to a wide range of behaviors. Below is an explanation of how plasma behaves in space:

1.3.1. Magnetic Fields and Electromagnetic Interactions

Magnetization: In space, plasma is highly influenced by magnetic fields. Since charged particles (electrons and ions) in plasma are sensitive to electromagnetic forces, they move in spirals or follow helical paths around magnetic field lines. This is called **magnetization** of the plasma.

- **Magnetic Confinement:** In regions like the Earth's **magnetosphere**, the plasma is confined by the planet's magnetic field, forming structures such as the **Van Allen radiation belts**, where high-energy particles are trapped.
- **Plasma Waves:** Interactions between charged particles and magnetic fields in space can produce **plasma waves**. These waves can accelerate particles to high speeds, contributing to phenomena like cosmic rays.

1.3.2. Solar Wind and Space Weather

Solar Wind: One of the most important plasma phenomena in space is the **solar wind**, a continuous stream of charged particles (mostly electrons and protons) flowing from the Sun's corona into space. The solar wind interacts with planetary magnetospheres, including Earth's, causing **geomagnetic storms** and **auroras**.

Coronal Mass Ejections (CMEs): Sometimes the Sun releases large bursts of plasma in events called CMEs. These bursts travel through space, and when they hit Earth's magnetosphere, they can disrupt satellite communications, cause power outages, and generate auroras.

1.3.3. Plasma Sheaths and Double Layers

Plasma Sheaths: When plasma encounters a solid object (e.g., a spacecraft or planetary surface), a **plasma sheath** forms around the object. This is a boundary layer where the plasma's electric fields and particle densities are altered. In space, these sheaths play a critical role in the behavior of spacecraft and satellites.

- **Double Layers:** In some plasma environments, **double**

layers form, which are regions with large electric fields where ions and electrons are separated. These are important in phenomena like auroras and shocks in space.

1.3.4. Space Plasmas Are Mostly Collisionless

In interplanetary and interstellar plasmas, the particle densities are so low that **collisions between particles** are rare. Unlike plasmas on Earth (such as in laboratory experiments), where particles frequently collide with one another, space plasmas interact primarily through electromagnetic forces. This leads to **collisionless shocks**—discontinuities in plasma properties like density, pressure, and magnetic field that are formed without particle collisions.

1.3.5. Kappa Distribution in Space Plasmas

Many space plasmas exhibit **non-Maxwellian distributions** of particle velocities, specifically **Kappa distributions**. In space, plasmas often exist in non-equilibrium states, where a higher proportion of particles have very high energies compared to what a Maxwell-Boltzmann distribution predicts. Kappa distributions are common in solar wind and other space plasmas, showing an extended tail of high-energy particles, which are responsible for cosmic rays and other energetic phenomena[6].

1.3.6. Plasma Behavior in Planetary Magnetospheres

Planets with magnetic fields, such as Earth and Jupiter, have **magnetospheres** that interact with solar wind plasma. The magnetic field of a planet shields it from solar and cosmic radiation by deflecting charged particles around the planet.

- **Auroras:** One visible effect of plasma-magnetic field interactions is the auroras (northern and southern lights) that occur near the magnetic poles when high-energy solar wind particles collide with atoms in the Earth's atmosphere, exciting them and causing them to emit light.

1.3.7. Plasma in Stars

Stellar Plasmas: The Sun and other stars are made of

plasma. The plasma in stars is incredibly hot and dense, and it is in a state of continuous nuclear fusion. This process releases vast amounts of energy, causing the outward pressure that balances the inward pull of gravity, maintaining the star's stability.

Plasma Dynamics in the Solar Corona: The solar corona (the outer layer of the Sun) is composed of plasma that is heated to millions of degrees. The high temperatures and magnetic fields in the corona generate phenomena such as solar flares, which are bursts of high-energy radiation and plasma.

1.3.8. Interstellar and Intergalactic Plasma

Interstellar Plasma: Between stars, there exists a diffuse plasma that fills the **interstellar medium**. This plasma can carry magnetic fields over vast distances and play a role in star formation, as gravitational collapse can concentrate plasma into dense regions that eventually form stars.

Cosmic Plasmas: On even larger scales, plasma pervades the space between galaxies in the form of the **intergalactic medium**. This plasma is tenuous but crucial for understanding large-scale structure formation in the universe.

1.3.9. Plasma Instabilities

Space plasmas are often subject to **plasma instabilities**, where small disturbances can grow and produce large-scale changes in plasma behavior. For example, the **Kelvin-Helmholtz instability** occurs when there is a velocity difference between two plasmas, such as the boundary between the solar wind and Earth's magnetosphere. This can create large, swirling structures in the plasma.

Source: Author

Figure 1. plasma propulsion systems, such as ion thrusters in space.

1.3.10. Plasma Propulsion

In space exploration, plasma behavior is also critical for **plasma propulsion systems**, such as **ion thrusters**. These systems generate plasma and use electric or magnetic fields to accelerate ions, producing thrust for spacecraft propulsion.

In space, plasma behaves in complex and dynamic ways, driven by electromagnetic forces, interactions with magnetic fields, and the unique low-collision environments. These behaviors are crucial to understanding space weather, planetary protection, the behavior of stars, and large-scale cosmic structures. Space plasmas are highly dynamic and can range from gentle flows like the solar wind to energetic and violent processes such as solar flares and interstellar shocks.

1.4. Plasma, Light and Lightling

The relationship between light, lightning, and plasma can be understood by exploring their physical properties and how they interact in nature. Here's an explanation of each and their connection:

Lightning: Lightning is a powerful electrical discharge caused by imbalances between storm clouds and the ground, or within the clouds themselves. When the potential difference becomes large enough, it causes the air to break down and become ionized, allowing electricity to flow. This ionized path is what we see as a lightning bolt.

Plasma: Plasma is often referred to as the fourth state of matter (besides solid, liquid, and gas). It consists of charged particles—free electrons and ions. Plasma is created when a gas is subjected to extreme energy, such as high temperatures or intense electric fields, causing the atoms to lose their electrons. In lightning, the extreme energy causes the air to become ionized, turning it into plasma.

Light: When lightning strikes, the intense heat (up to 30,000 K) ionizes the air and creates plasma, which emits light. This light comes from the excited atoms and molecules in the plasma as they return to their normal states after being energized. The visible flash we see during a lightning strike is a result of the light emitted by the plasma formed in the air.

In summary, **lightning creates plasma by ionizing the air through high-energy electrical discharges**. The plasma, in turn, emits light as a result of the excitation of particles. Thus, lightning, light, and plasma are interconnected phenomena resulting from intense energy in the atmosphere.

1.4.1. Syngas production from lightning

The possibility of lightning producing syngas (a mixture of hydrogen and carbon monoxide, primarily) in the atmosphere is an interesting concept, but it is unlikely under normal atmospheric conditions. Here's why:

1. **Syngas Production**: Syngas is typically produced through processes like **gasification** or **steam reforming**, which involve the breakdown of hydrocarbons (such as coal, biomass, or natural gas) at high temperatures and in the presence of a controlled amount of oxygen or steam. These processes require both carbon-containing material and hydrogen sources, as well as specific catalysts and

conditions.

2. **Lightning's Energy**: Lightning releases an immense amount of energy, with temperatures reaching up to 30,000 K. While this energy is sufficient to ionize gases in the air (producing plasma) and even initiate chemical reactions, the atmosphere mostly consists of nitrogen, oxygen, water vapor, and trace gases. It doesn't naturally contain significant amounts of hydrocarbons, which are required to produce syngas. In order for syngas to form, there would need to be a source of carbon (such as methane or other organic compounds) and a process to strip oxygen atoms from carbon dioxide or water molecules to form carbon monoxide and hydrogen.
3. **Possible Reactions**: While lightning can trigger chemical reactions in the atmosphere (such as the formation of **ozone** from oxygen or **nitrogen oxides** from nitrogen), the conditions required for syngas production—namely, the presence of hydrocarbon fuels and appropriate reaction conditions—are not typically found in the atmosphere. The energy from lightning alone, without a carbon-rich source, wouldn't lead to syngas formation.
4. **Terrestrial Sources**: In some rare cases, if lightning were to strike a carbon-rich source like biomass (e.g., in a forest fire), some syngas might theoretically form as a byproduct of the high temperatures breaking down the organic material. But even in these cases, the production of syngas would be minimal compared to the complex array of other combustion products, such as carbon dioxide, carbon monoxide, water vapor, and other hydrocarbons.

Conclusion: Lightning can cause intense chemical reactions in the atmosphere, but the specific conditions needed to produce syngas—carbon-rich materials and controlled oxygen input—are not typically present in the atmosphere. Therefore, it's unlikely that significant amounts of syngas would be produced solely by lightning under normal environmental conditions.

Source: Author

Figure 2. The image of lightning as a powerful electrical discharge, forming a plasma channel.

1.4.2. Lightning and air neutralization

an interesting point about the role of lightning in atmospheric chemistry and its potential contribution to air neutralization. While it's true that lightning can influence certain atmospheric processes, its role in "neutralizing" the air is more complex and may be misunderstood if we don't carefully consider how it fits into the larger context of atmospheric dynamics. Let's break this down:

a. Lightning's Production of Nitrogen Compounds:

Lightning produces **nitrogen oxides (NO and NO_2)** through the intense heating of air, which breaks apart nitrogen (N_2) and oxygen (O_2) molecules, allowing them to combine. These nitrogen oxides are precursors to **nitric acid (HNO_3)** when they react with water vapor in the atmosphere, which can lead to **acid rain**.

However, nitrogen oxides are also important because they

contribute to the formation of **nitrates**, which are essential nutrients for plants when deposited on land. This can have a beneficial effect on ecosystems, particularly in areas where nitrogen is a limiting nutrient.

b. Ozone Production:

Lightning can also produce **ozone (O_3)**, particularly in the upper atmosphere (stratosphere). Ozone in the stratosphere plays a critical role in absorbing harmful ultraviolet (UV) radiation from the sun. However, in the lower atmosphere (troposphere), ozone can act as an air pollutant, contributing to smog formation and respiratory problems.

In certain cases, ozone can react with pollutants in the air (such as volatile organic compounds or methane), helping to oxidize or break them down, thus contributing to some degree of "cleaning" the air.

c. Cleaning or Neutralizing the Air?:

The term "neutralization" typically refers to the process of reducing acidity or balancing chemical compositions. Lightning can contribute to atmospheric chemistry by:

- Producing **ozone** that can help break down certain air pollutants.
- Generating **nitrogen oxides**, which can lead to the creation of **nitrates** that are beneficial for plant growth.

However, **lightning does not neutralize acids** in the atmosphere. In fact, the nitrogen oxides produced by lightning often contribute to **acid rain** formation when they transform into nitric acid (HNO_3). So, while lightning may create conditions that support air "cleaning" in some ways, it does not function as a significant neutralizing force in terms of balancing acid rain or overall atmospheric acidity.

d. Contribution to Overall Atmospheric Chemistry:

Lightning plays a role in the natural nitrogen cycle by fixing nitrogen from the atmosphere and converting it into nitrogen compounds that can eventually be deposited in soils as nutrients.

This is an important contribution to ecosystems, particularly in regions where industrial emissions are lower, and natural processes dominate.

However, when considering **air neutralization on a global scale**, lightning's contribution is relatively small compared to other natural processes (like soil buffering or plant uptake of atmospheric compounds) and anthropogenic efforts (like reducing industrial emissions of pollutants). The amount of nitrogen oxides produced by lightning is far smaller compared to human-made sources, such as combustion engines and industrial activities.

e. Conclusion:

Lightning does contribute to atmospheric processes and can have localized effects on air chemistry, particularly through the production of ozone and nitrogen compounds. However, its role as a major contributor to **air neutralization** is limited. In fact, lightning's production of nitrogen oxides often contributes to the formation of nitric acid and, subsequently, **acid rain**. While it is a fascinating and important natural phenomenon, its capacity for neutralizing the atmosphere is overshadowed by its role in forming reactive nitrogen compounds and its relatively small-scale impact compared to human activities.

So, while lightning does contribute to atmospheric chemistry in notable ways, its overall impact on air neutralization is not as significant as other natural or anthropogenic processes.

1.4.3. Lightning contributions to humans and the environment

Lightning has several direct and indirect contributions to humans and the environment, both positive and negative. Here's a summary of its **benefits and contributions**:

a. Nitrogen Fixation and Fertility:

One of the most important benefits of lightning is its role in **nitrogen fixation**:

- **Nitrogen oxides (NO_x)** formed by lightning combine with rainwater to produce **nitrates (NO_3^-)**, which are deposited into the soil. These nitrates are essential nutrients for plant growth and are a crucial part of the **nitrogen cycle**.

- This natural process helps fertilize the land, especially in ecosystems like forests and grasslands where lightning strikes frequently but human agriculture and fertilizer use are minimal. This contribution to soil fertility supports plant life and ecosystems, indirectly benefiting humans by maintaining ecological balance.

b. Ozone Production:

- Lightning generates **ozone (O_3)**, particularly in the upper atmosphere (stratosphere), which plays a vital role in protecting life on Earth by absorbing harmful **ultraviolet (UV) radiation** from the Sun. This shielding effect is critical for preventing overexposure to UV rays, which can cause skin cancer and other health problems.
- However, ozone generated in the lower atmosphere (troposphere) can contribute to smog and respiratory issues, so the benefits are more relevant in the upper atmosphere.

c. Weather and Climate Influences:

- **Lightning** is a key component of thunderstorms and other natural weather events. These storms help redistribute heat in the atmosphere, influence global weather patterns, and even contribute to natural rainfall in many regions.
- In areas with regular lightning storms, rainfall supports **agricultural systems** by providing much-needed water for crops and local water cycles.

d. Forest Regeneration and Ecosystem Health:

- In certain ecosystems, like **savannas** and **pine forests**, lightning-induced fires play a role in maintaining biodiversity and promoting **forest regeneration**. Periodic fires can clear out dead vegetation, allowing new growth to flourish and preventing larger, more destructive fires from occurring later.

- This natural process supports ecosystems that are fire-adapted, helping them regenerate and maintain healthy cycles of growth.

e. Scientific and Technological Inspiration:

- **Research on lightning** has driven scientific advancements in understanding **electricity** and atmospheric processes. For instance, the study of electrical discharges in lightning has helped scientists understand more about electrical conduction and plasma physics, contributing to advances in **energy generation** and **electrical engineering**.
- Lightning detection and monitoring technologies are used in weather forecasting, aviation safety, and disaster management, helping reduce risks to human life and infrastructure.

f. Wildfire Risks (Negative Impact):

- Although not a direct benefit, it is important to note that lightning can have negative impacts by causing **wildfires,** especially in dry regions. While some wildfires are ecologically beneficial, uncontrolled large-scale fires can threaten human life, property, and air quality. Lightning is responsible for a significant percentage of natural wildfires.

g. Inspiration in Culture and Mythology:

- Throughout history, lightning has been a symbol of power and energy in many cultures. It has inspired myths, religious beliefs, and artistic representations, contributing to cultural development and human understanding of nature's forces.

h. Conclusion:

The **benefits and contributions of lightning to humans** can be categorized into several areas:

- **Ecological Contributions**: Lightning plays a key role in the nitrogen cycle, fertilizing soils and supporting plant life. It also helps maintain ecosystems through fire regeneration in fire-adapted areas.

- **Environmental Protection**: The formation of ozone in the upper atmosphere provides protection from harmful UV radiation.
- **Scientific Advancement**: The study of lightning has advanced our understanding of electricity, atmospheric processes, and led to improvements in technology for weather prediction and safety.
- **Cultural and Inspirational**: Lightning has inspired human culture, myths, and scientific curiosity.

While lightning poses risks, particularly through wildfires and the creation of pollutants like nitrogen oxides, its natural contributions to ecosystems and atmospheric dynamics are essential and beneficial to both the environment and humanity.

1.4.4. Plasma and Initiation of life

The idea that **plasma** could be considered an initiator of life, particularly in terms of providing the energy necessary for life's emergence, is a fascinating concept. While it's not a mainstream theory in the field of biology, plasma does hold some intriguing possibilities when viewed through the lens of energy and the origins of life.

Here's how plasma could potentially play a role in the **initiation of life**:

a. Plasma as a Source of Energy:

- **Energy is a critical requirement for life**. All known life forms rely on energy to drive the chemical reactions necessary for metabolism, growth, and reproduction. In Earth's early history, before the appearance of life, there were various sources of energy that could have catalyzed the formation of complex organic molecules from simpler chemicals.
- Plasma, as an ionized gas consisting of free electrons and

ions, is highly energetic. It can be generated in a variety of ways, such as through **lightning**, solar flares, or cosmic radiation. Plasma creates a highly energetic environment that could trigger chemical reactions in a prebiotic setting.

b. Plasma and Early Earth Conditions:

- In the **early Earth**, lightning storms were frequent, and the atmosphere was rich in gases like methane, ammonia, hydrogen, and water vapor. Lightning and the plasma it generated could have provided the energy needed to catalyze the formation of **organic molecules**, as suggested by the famous **Miller-Urey experiment** in the 1950s. In that experiment, simulating early Earth's conditions with electrical sparks led to the creation of amino acids, which are the building blocks of life.
- Plasma could play a similar role by interacting with gases and water in Earth's early atmosphere or oceans, providing the energy necessary for complex molecules like amino acids, nucleotides, and lipids to form.

c. Plasma's Ability to Break and Reform Chemical Bonds:

- Plasma, due to its energetic nature, is capable of **breaking chemical bonds** and creating **free radicals**—highly reactive molecules that can interact with other atoms and molecules to form new compounds. This ability to break and reform bonds is essential for creating the types of complex organic molecules that could serve as precursors to life.
- **Free radicals** produced by plasma could have accelerated the formation of more complex organic compounds, including amino acids and nucleotides, which are fundamental to life.

d. Plasma and the Formation of Membranes:

- Life, as we know it, is dependent on the formation of **cell membranes** to create distinct boundaries between the inside and outside of cells. Plasma could theoretically play a role in promoting the formation of lipid bilayers or other structures that serve as primitive cell membranes.

- Plasma-generated energy might help assemble these early membrane structures from simple molecules, helping to compartmentalize reactions and eventually lead to the formation of the first protocells.

e. Plasma in Space: The Role of Cosmic Plasmas:

- **Plasma** is abundant in space, where stars, including our Sun, are massive sources of plasma energy. Some theories suggest that life on Earth could have been influenced or seeded by **cosmic events**, including the delivery of organic compounds to Earth via comets or meteorites, which are bombarded by cosmic plasma.
- The interaction of these space-based plasma environments with organic molecules in space or during their entry into Earth's atmosphere could further contribute to the theory that plasma has a role in initiating life, possibly by altering organic molecules to make them more reactive or forming complex compounds.

f. Plasma and Extremophiles: Clues from Life at the Edge:

- The discovery of **extremophiles**, organisms that thrive in extreme conditions, such as high radiation or near plasma-rich environments (like around hydrothermal vents), suggests that life can survive and even thrive in conditions similar to those created by plasma. This hints that the highly energetic environments created by plasma discharges may not be as hostile to life as previously thought.
- These organisms provide insights into how life might have evolved under extreme conditions, suggesting that plasma-rich environments could have played a role in fostering the emergence of early life forms capable of adapting to high-energy settings.

g. Potential for Synthetic Biology:

- In modern science, researchers are using plasma in **synthetic biology** and **biotechnology** to create new biomaterials and influence biological processes. **Cold plasma technology** is already being used in medicine for wound healing and sterilization due to its ability to manipulate biological materials without damaging them.

- This potential shows that plasma can interact with biological systems in a controlled way, possibly offering insights into how plasma could have been part of the processes that led to life in the first place.

In general, wWhile plasma may not be the direct cause of life, it could certainly be seen as a **catalyst** or **initiator** for the energy-intensive reactions that led to the formation of life's building blocks on early Earth. The high-energy environment created by plasma could have facilitated the **formation of complex organic molecules**, breaking and reforming bonds in ways that are essential to the origin of life.

In this sense, plasma serves as a powerful and intriguing source of energy, capable of driving the chemical transformations necessary for the emergence of life. Whether on early Earth, through lightning storms, or in cosmic environments, plasma's contribution to the initiation of life, particularly as a source of energy and molecular transformation, is an exciting area of study that could deepen our understanding of how life began.

1.4.5. Plasma as "Light Upon Light"

The phrase **"Light upon Light"** from **Qur'an 24:35** has been interpreted in various ways throughout history, particularly within Islamic scholarship and mysticism. It is part of the verse known as the **"Ayat al-Nur" (Verse of Light)**, which has a deeply symbolic and metaphysical significance, often understood as describing the divine light of guidance and knowledge. While classical interpretations focus on the spiritual and metaphysical meanings, considering **plasma** as a possible analogy for "Light upon Light" is an intriguing and modern perspective, especially given plasma's unique characteristics in the physical universe.

Plasma as "Light Upon Light". If we contemplate the concept of **plasma** in the context of the phrase "Light upon Light," several interesting parallels emerge:

a. Plasma's Nature as Light-Emitting:

Plasma is often described as the **fourth state of matter**, and it is highly energetic, made up of charged particles (ions and electrons). When these particles move and interact, they emit **light** across a broad spectrum. Plasma is responsible for some of the most spectacular displays of light in nature, such as:

- **Stars**, including the Sun, are composed of plasma, and they emit light as a result of nuclear reactions within.
- **Auroras** (Northern and Southern Lights) are a plasma phenomenon, where solar wind interacts with Earth's magnetic field to produce glowing light in the upper atmosphere.

This idea of **light being produced by plasma** can be linked metaphorically to the "Light upon Light" phrase, where the continuous emission and interaction of energy in plasma could be seen as layers of light manifesting upon each other.

b. Infinite Layers of Light in Plasma:

Plasma's interaction with energy can be seen as an ongoing and **continuous generation of light**. In the case of **"Light upon Light"**, this could be understood as an infinite layering or cascading of illumination, where one form of light leads to another, much like plasma's dynamic and glowing nature.

- In stars, for instance, nuclear fusion produces layers of energy and light that radiate outward in cycles, continuously illuminating and sustaining the universe.
- Plasma can also produce **different wavelengths of light**, from visible light to ultraviolet, depending on its energy and environment, much like the concept of **divine light** manifesting in different forms of guidance, knowledge, or spiritual realization.

c. Plasma's Universal Presence:

Plasma is the most abundant form of matter in the **universe**, making up stars, the interstellar medium, and even much of the ionosphere. In this way, plasma surrounds us and is part of the fabric of the universe, just as divine light, symbolized by **"Light upon Light"**, is often interpreted to permeate all of creation.

- Plasma's pervasive nature in the cosmos echoes the idea of divine light as **all-encompassing**, illuminating not just the physical realm but the spiritual one as well.

d. Energy and Guidance:

The **Qur'anic verse** focuses on **God's light as a guiding force**—guiding people out of darkness and confusion toward truth and understanding. Plasma, in a physical sense, plays a role in illuminating and providing energy to sustain life (through the Sun) and drive many processes on Earth.

- Just as **plasma** provides light and energy, the **divine light** described in the Qur'an is a metaphor for spiritual energy and **guidance** that leads to understanding and enlightenment.

e. Plasma as a Symbol of Creation:

In cosmology, plasma is thought to have been the original state of matter shortly after the **Big Bang**, before cooling into gas and eventually forming the universe as we know it. This state of primal plasma could symbolize the **primordial light** of creation, aligning with mystical interpretations of "Light upon Light" as referring to God's creative power, continuously manifesting throughout existence.

- In this sense, plasma could represent **God's first act of creation**, where divine energy forms and sustains the material universe.

f. Metaphysical Interpretations:

In Islamic mysticism (**Sufism**), the concept of **light** is often used metaphorically to describe spiritual **illumination, knowledge, and divine presence**. Plasma, which is light in its most energetic and dynamic form, could serve as a modern metaphor for this divine light:

- Plasma's constant movement, interaction, and emission of

light could symbolize the **infinite** and ever-present **flow of divine wisdom**, much like how **Sufi** interpretations describe "Light upon Light" as an endless deepening of spiritual awareness and proximity to the divine.

g. Conclusion:

While the classical interpretation of **"Light upon Light"** in the Qur'an (24:35) centers around divine light, guidance, and spiritual illumination, there is an interesting **scientific and metaphorical parallel** to be drawn with **plasma**. Plasma, being a highly energetic and light-emitting state of matter, shares many characteristics with the idea of **light as a dynamic, infinite force**. It is the foundation of much of the universe's visible light and energy, just as divine light is considered the foundation of truth and spiritual understanding in the Islamic tradition.

Considering plasma in this context is a modern way to reflect on the Qur'anic verse, linking the **cosmic** and **spiritual dimensions** of light, and demonstrating how ancient wisdom can find resonance in modern scientific concepts. While not a traditional interpretation, the metaphor of plasma as **"Light upon Light"** can deepen our understanding of both **science** and **spirituality** by highlighting the connection between the physical and the metaphysical realms.

2: MATHEMATICAL FOUNDATIONS OF PLASMA MODELLING

The mathematical foundations of plasma modelling are critical for understanding and predicting the behavior of plasmas in various environments. This chapter delves into the fundamental equations, kinetic theory, and plasma approximation methods that form the basis of plasma modelling.

2.1 Fundamental Equations

Plasma behaviour is governed by a set of fundamental equations that describe the dynamics of charged particles and their interactions with electromagnetic fields. The primary equations used in plasma modelling include Maxwell's equations and the fluid equations.

2.1.1. Maxwell's Equations

These equations describe how electric and magnetic fields propagate and interact with charges and currents in the plasma. They are expressed as:

- Gauss's Law for Electricity:

$$\nabla \cdot \mathbf{E} = \frac{\rho}{\epsilon_0}$$

where **E** is the electric field, ρ is the charge density, and ϵ_0 is the permittivity of free space [8].

- Gauss's Law for Magnetism:

$$\nabla \cdot \mathbf{B} = 0$$

where **B** is the magnetic field [9].

- Faraday's Law of Induction:

$$\nabla \times \mathbf{E} = -\partial \mathbf{B}/\partial t$$

This formula indicating that a time-varying magnetic field generates an electric field [10].

- Ampère's Law (with Maxwell's correction):

$$\nabla \times \mathbf{B} = \mu_0 \mathbf{J} + \mu_0 \epsilon_0 \frac{\partial \mathbf{E}}{\partial t}$$

where $\boldsymbol{J}$ is the current density and $\boldsymbol{\mu}_0$ is the permeability of free space [11].

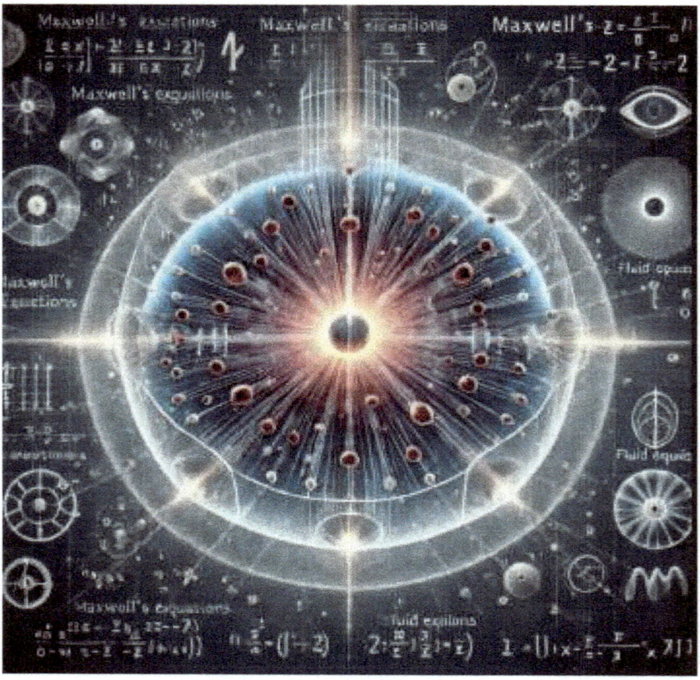

Source: Author

Figure 3. The schematic image showing how plasma behavior is governed by fundamental equations like Maxwell's equations and fluid equations

These equations are coupled with the Lorentz force law, which describes the force on a charged particle due to electric and magnetic fields:

$$\mathbf{F} = q(\mathbf{E} + \mathbf{v} \times \mathbf{B})$$

where $\boldsymbol{q}$ is the charge of the particle and $\boldsymbol{v}$ is its velocity [12].

2.1.2. Fluid Equations

Plasmas can also be described using fluid models, which

treat the plasma as a continuous medium rather than a collection of individual particles. The fluid equations consist of the continuity equation, momentum equation, and energy equation.

- Continuity Equation:

$$\frac{\partial n}{\partial t} + \nabla \cdot (n\mathbf{v}) = 0$$

where ***n*** is the particle density and ***v*** is the fluid velocity. This equation ensures mass conservation[13].

2.1.3. Momentum Equation (Navier-Stokes Equation for Plasmas)

$$m\left(\frac{\partial \mathbf{v}}{\partial t} + \mathbf{v} \cdot \nabla \mathbf{v}\right) = q(\mathbf{E} + \mathbf{v} \times \mathbf{B}) - \nabla p$$

where ***m*** is the mass of the particles and ***p*** is the pressure. This equation describes the momentum conservation and includes forces due to electric and magnetic fields, as well as pressure gradients[14].

2.1.4. Energy Equation:

The energy conservation can be formulated as:

$$\frac{\partial \epsilon}{\partial t} + \nabla \cdot (\epsilon \mathbf{v}) = -\nabla \cdot \mathbf{q} + Q$$

where ϵ is the energy density, ***q*** is the heat flux, and ***Q*** represents energy sources or sinks. This equation ensures energy conservation [15].

2.2. Kinetic Theory

Kinetic theory provides a more detailed description of plasma by considering the distribution of individual particle velocities. The primary tool in kinetic theory is the Boltzmann equation, which describes the evolution of the distribution function ***f(x,v,t)*** that gives the number density of particles in phase space.

2.2.1. Boltzmann Equation:

$$\frac{\partial f}{\partial t} + \mathbf{v} \cdot \nabla_{\mathbf{x}} f + \frac{\mathbf{F}}{m} \cdot \nabla_{\mathbf{v}} f = \left(\frac{\partial f}{\partial t}\right)_{\text{coll}}$$

where ∇_x and ∇_v are the spatial and velocity gradients, respectively, and $(\partial f/\partial t)_{coll}$ represents the collision term, accounting for particle interactions [16].

2.2.2 Distribution Functions and Moments:

The distribution function $f(x,v,t)$ provides a comprehensive description of the plasma state. Moments of the distribution function yield macroscopic quantities:

- Density:

  $n(x,t)= \int f(x,v,t)\, d^3\mathbf{v}$

- Velocity:

  $v(x,t)=1n\int vf(x,v,t)\, d^3v$

- Pressure Tensor:

  $P(x,t)=m\int(v-u)(v-u)f(x,v,t)\, d^3v$

  where **u** is the mean velocity [17].

2.3 Plasma Approximation Methods

Several approximation methods are employed in plasma modelling to simplify the complex equations and make them tractable for analytical and numerical solutions. The primary methods include magnetohydrodynamics (MHD), two-fluid and multi-fluid models, and particle-in-cell (PIC) methods.

Magnetohydrodynamics (MHD): MHD is a macroscopic model that treats the plasma as a single conducting fluid interacting with electromagnetic fields. The MHD equations combine the fluid equations with Maxwell's equations, assuming that the plasma is sufficiently collisional to be described by fluid dynamics.

2.3.1. Ideal MHD:

Assumes the plasma is perfectly conducting, neglecting resistivity:

$$\frac{\partial \rho}{\partial t} + \nabla \cdot (\rho \mathbf{v}) = 0$$
$$\rho \left(\frac{\partial \mathbf{v}}{\partial t} + \mathbf{v} \cdot \nabla \mathbf{v} \right) = -\nabla p + \mathbf{J} \times \mathbf{B}$$
$$\frac{\partial \mathbf{B}}{\partial t} = \nabla \times (\mathbf{v} \times \mathbf{B})$$

where $\boldsymbol{\rho}$ is the mass density, **J** is the current density, and **B** is the magnetic field [18].

2.3.2. Resistive MHD

Resistive MHD Includes the effects of finite resistivity:

$$\mathbf{E} + \mathbf{v} \times \mathbf{B} = \eta \mathbf{J}$$

where $\boldsymbol{\eta}$ is the resistivity [19].

2.3.3. Two-Fluid and Multi-Fluid Models

These models consider electrons and ions as separate fluids, each with its own set of fluid equations. This approach allows for a more detailed description of plasma behavior, particularly for phenomena where electron and ion dynamics differ significantly.

Two-Fluid Model: Describes the dynamics of electrons and ions separately:

$$m_e n_e \left(\frac{\partial \mathbf{v}_e}{\partial t} + \mathbf{v}_e \cdot \nabla \mathbf{v}_e \right) = -e n_e (\mathbf{E} + \mathbf{v}_e \times \mathbf{B}) - \nabla p_e + \mathbf{R}_{ei}$$
$$m_i n_i \left(\frac{\partial \mathbf{v}_i}{\partial t} + \mathbf{v}_i \cdot \nabla \mathbf{v}_i \right) = e n_i (\mathbf{E} + \mathbf{v}_i \times \mathbf{B}) - \nabla p_i + \mathbf{R}_{ie}$$

where m_e and m_i are the electron and ion masses, n_e and n_i are the electron and ion densities, $\mathbf{v}_e$ and $\mathbf{v}_i$ are the electron and ion velocities, p_e and p_i are the electron and ion pressures, and $\mathbf{R}_{ei}$ and $\mathbf{R}_{ie}$ are the electron-ion collision terms [20].

2.3.4. Particle-in-Cell (PIC) Methods

PIC methods are kinetic models that simulate the behavior of individual particles in the plasma. In PIC simulations, particles are represented by computational particles, each carrying a portion of the charge and mass. The equations of motion for these particles are solved using the Lorentz force law, while the electromagnetic fields are updated on a grid.

Equations of Motion:

$$\frac{d\mathbf{x}_i}{dt} = \mathbf{v}_i$$
$$\frac{d\mathbf{v}_i}{dt} = \frac{q_i}{m_i}(\mathbf{E}(\mathbf{x}_i) + \mathbf{v}_i \times \mathbf{B}(\mathbf{x}_i))$$

where $\mathbf{x}_i$ and $\mathbf{v}_i$ are the position and velocity of the iiith particle, and q_i and m_i are its charge and mass [21].

2.3.5. Field Update:

Maxwell's equations are solved on a grid to update the electric and magnetic fields:

$$\nabla\cdot\mathbf{E}=\rho/\epsilon_0$$

$$\nabla\cdot\mathbf{B}=0$$

$$\nabla\times\mathbf{E}=-(\partial t/\partial\mathbf{B})$$

$$\nabla\times\mathbf{B}=\mu_0\mathbf{J}+\mu_0\epsilon_0(\partial E/\partial t)$$

where ρ and $\mathbf{J}$ are the charge and current densities obtained from the particle positions and velocities [22].

In summary, Table 1. organizes various plasma models by the equations they use and the underlying physical laws, helping provide an overview of the complexity and scope of plasma modeling across different applications. This table has two main key components:

- **Equations**: This column lists the primary mathematical models used in each type of plasma simulation.
- **Governing Physics Laws**: The key physics laws and principles that define each plasma model, including electromagnetism, fluid dynamics, thermodynamics, and quantum mechanics.

2.4 Applications of Plasma Modelling

Plasma modelling is pivotal in numerous scientific and engineering applications. Here, we highlight some key areas where these models are extensively used.

Nuclear Fusion: One of the most prominent applications of plasma modelling is in the field of nuclear fusion. Controlled thermonuclear fusion aims to replicate the energy production process of the sun on Earth, offering a potential source of clean and nearly limitless energy. Magnetic confinement devices like tokamaks and stellarators rely on plasma modelling to understand and control the plasma behavior, optimize confinement, and prevent instabilities [23].

Space Plasmas: Plasma modelling plays a crucial role in space science, helping to understand phenomena such as the solar wind, Earth's magnetosphere, and cosmic rays. Models of space plasmas are used to predict space weather, which can affect satellite operations, communication systems, and power grids on Earth [24].

Table 1. Table organizes various plasma models by the equations they use and the underlying physical laws,

Plasma Model	Equations	Governing Physics Laws
Particle-in-Cell (PIC)	Maxwell's Equations, Newton's Second Law	**Electromagnetism** (Maxwell's Laws), **Kinetics** (Newton's Laws)
Fluid Plasma Models	Navier-Stokes Equations, Continuity Equation	**Conservation Laws** (mass, momentum, energy), **Hydrodynamics**
Magnetohydrodynamics (MHD)	MHD Equations (coupled fluid and electromagnetic)	**Electromagnetism** (Maxwell's Equations), **Hydrodynamics**
Gyrokinetic Models	Gyrokinetic Equation, Poisson's Equation	**Kinetics** (Vlasov Equation), **Electromagnetism**
Hybrid Particle-MHD	Particle Vlasov Equation, MHD Equations	**Kinetic Theory** (Particle Motion), **Fluid Dynamics**, **Electromagnetism**
Plasma Edge Fluid Models	Braginskii Equations, Energy Balance Equation	**Thermodynamics**, **Hydrodynamics** (Transport Processes)
Quantum Hydrodynamics (QHD)	Quantum Hydrodynamics Equations	**Quantum Mechanics**, **Fluid Dynamics**
Fokker-Planck Models	Fokker-Planck Equation, Langevin Equation	**Statistical Mechanics**, **Kinetic Theory**
Plasma Kinetic Models	Boltzmann Equation, Collision Term	**Kinetic Theory**, **Statistical Physics** (Boltzmann Transport)
Plasma Turbulence Models	Gyrokinetic Turbulence Equations, Navier-Stokes	**Turbulence Theory**, **Fluid Dynamics**, **Electromagnetism**
Plasma-Material Interaction	Particle-in-Cell (PIC), Surface Reaction Models	**Surface Chemistry**, **Kinetic Theory**, **Fluid Dynamics**
Cold Plasma Models	Drift-Diffusion Equations, Poisson's Equation	**Electromagnetism**, **Kinetics** (Drift-Diffusion)
Radiative Plasma Models	Radiative Transfer Equations, Energy Conservation	**Radiative Transport**, **Thermodynamics**, **Quantum Mechanics**

Source: Author

Industrial Applications: In industries, plasmas are used for material processing techniques such as plasma etching, deposition, and surface treatment. Plasma modelling helps optimize these processes, improving efficiency and product quality. For instance, in semiconductor manufacturing, precise control of plasma processes is essential for creating intricate microelectronic circuits [25].

Environmental Applications: Plasmas are also employed in environmental technologies for pollution control and waste treatment. Plasma-based systems can break down pollutants in air and water, converting harmful substances into less toxic forms. Modelling these processes ensures effective and energy-efficient operation [26].

Medical Applications: In the medical field, non-thermal plasmas are used for sterilization, wound healing, and cancer treatment. Plasma modelling contributes to understanding the interactions between plasmas and biological tissues, leading to safer and more effective medical devices and therapies [27].

Conclusion

Understanding the mathematical foundations of plasma modelling is essential for advancing research and applications in this dynamic field. The fundamental equations, kinetic theory, and various approximation methods provide the tools necessary to describe and predict plasma behavior under different conditions. As computational power continues to grow and new numerical techniques are developed, plasma modelling will become even more sophisticated, driving innovations in energy, space science, industry, environment, and medicine.

3: NUMERICAL METHODS IN PLASMA MODELLING

Numerical methods are essential for solving the complex equations that describe plasma behavior. Given the highly nonlinear and multidimensional nature of these equations, analytical solutions are often impractical or impossible. This chapter provides a detailed overview of the numerical methods used in plasma modelling, including discretization techniques, solving linear and nonlinear equations, and addressing issues of stability and convergence.

3.1 Discretization Techniques

Discretization is the process of transforming continuous equations into discrete forms that can be solved numerically. This involves approximating the equations on a grid or mesh that spans the spatial domain of interest. The primary discretization techniques used in plasma modelling include finite difference, finite element, and finite volume methods.

3.1.1. Finite Difference Methods (FDM)

The finite difference method approximates derivatives by differences between function values at discrete points. For example, the first derivative of a function *u(x)* can be approximated as:

$$\frac{du}{dx} \approx \frac{u(x+\Delta x) - u(x)}{\Delta x}$$

where Δx is the grid spacing. Higher-order derivatives can be approximated similarly using central, forward, or backward difference formulas [28].

FDM is straightforward to implement and is widely used for problems with simple geometries. However, its accuracy depends

on the grid resolution, and it can be challenging to apply to complex boundary conditions or irregular domains [29].

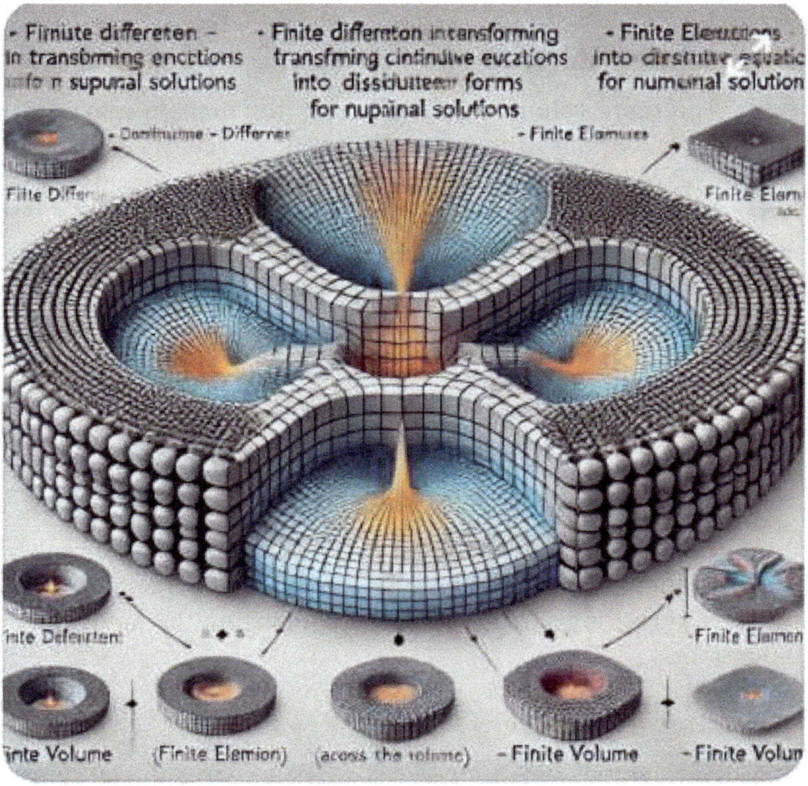

Source: Author

Figure 4. the schematic image illustrating the concept of discretization and the primary techniques used in plasma modeling

3.1.2. Finite Element Methods (FEM):

The finite element method divides the domain into smaller, simpler pieces called elements. The solution is approximated by piecewise polynomial functions within each element. This approach is highly flexible and can handle complex geometries and boundary conditions. The weak form of the differential equations is derived using weighted residual methods, and the resulting system of equations is solved for the nodal values [30].

FEM is particularly powerful for problems involving

irregular domains and varying material properties. It is widely used in structural mechanics, fluid dynamics, and electromagnetic simulations [31].

3.1.3. **Finite Volume Methods (FVM):**

The finite volume method divides the domain into control volumes and applies the integral form of the conservation laws. The fluxes of conserved quantities (e.g., mass, momentum, energy) are calculated across the boundaries of each control volume. This ensures conservation at a discrete level and is particularly suited for problems involving conservation laws and shock waves [32].

FVM is commonly used in computational fluid dynamics (CFD) and is known for its robustness and conservation properties. It can handle complex geometries and is well-suited for compressible flow simulations [33].

3.2 Solving Linear and Nonlinear Equations

Once the equations are discretized, the next step is solving the resulting system of linear or nonlinear equations. Various numerical methods are employed depending on the nature of the equations.

3.2.1. Iterative Solvers for Linear Systems

Iterative methods are preferred for large, sparse linear systems commonly arising in plasma modelling. Some popular iterative solvers include:

Jacobi Method: This method iteratively updates the solution by solving the diagonal elements separately:

$$x_i^{(k+1)} = \frac{1}{a_{ii}} \left(b_i - \sum_{j \neq i} a_{ij} x_j^{(k)} \right)$$

where a_{ij} are the coefficients of the matrix, b_i are the elements of the right-hand side vector, and $x_i^{(k)}$ is the solution at iteration k [34].

Gauss-Seidel Method: An improvement over the Jacobi method, it uses the latest available values during iteration:

$$x_i^{(k+1)} = \frac{1}{a_{ii}}\left(b_i - \sum_{j<i} a_{ij}x_j^{(k+1)} - \sum_{j>i} a_{ij}x_j^{(k)}\right)$$

This typically converges faster than the Jacobi method [35]

Conjugate Gradient Method: Suitable for symmetric, positive-definite matrices, it minimizes the residual error iteratively in the direction of the conjugate gradient

$$x^{(k+1)} = x^{(k)} + \alpha_k p^{(k)}$$

$$p^{(k+1)} = r^{(k+1)} + \beta_k p^{(k)}$$

where $\boldsymbol{\alpha_k}$ alpha_kαk and β_k are determined to minimize the error, $p^{(k)}$ is the search direction, and $r^{(k+1)}$ is the residual [36].

3.2.2. Direct Solvers for Linear Systems

Direct methods, such as LU decomposition, solve the system by factorizing the matrix into lower and upper triangular matrices. While accurate, they are computationally expensive and less suited for large systems [37].

- **Nonlinear Solvers**: Many plasma problems lead to nonlinear equations. Newton-Raphson and quasi-Newton methods are commonly used to solve these systems.
- **Newton-Raphson Method**: Iteratively refines the solution by linearizing the nonlinear equations:
- $x^{(k+1)}=x^{(k)}-J^{-1}(x^{(k)})F(x^{(k)})$

  where $\boldsymbol{J}$ is the Jacobian matrix of partial derivatives and $\boldsymbol{F}$ is the vector of nonlinear equations [38]
- **Quasi-Newton Methods:** These methods approximate the Jacobian matrix to reduce computational effort. The Broyden's method is a popular choice [39].

3.3 Stability and Convergence

Ensuring stability and convergence is crucial for the reliability of numerical simulations. Stability refers to the behavior of numerical errors over time, while convergence ensures that the numerical solution approaches the true solution as the grid is refined.

3.3.1. Time-Stepping Methods:

For time-dependent problems, explicit and implicit time-stepping methods are used.

- **Explicit Methods**: These methods compute the solution at the next time step using the current values. They are easy to implement but require small time steps for stability. The Forward Euler method is a simple example:
- $u^{n+1}=u^n+\Delta t f(u^n)$

  where u^n is the solution at time step n and Δt is the time step size [40].

  - **Implicit Methods**: These methods involve solving a system of equations at each time step, allowing for larger time steps. The Backward Euler method is an example:

    $u^{n+1}=u^n+\Delta t f(u^{n+1})$

  Implicit methods are more stable but computationally intensive [41].

3.3.2. CFL Condition:

The Courant-Friedrichs-Lewy (CFL) condition is a necessary criterion for stability in explicit time-stepping methods. It ensures that the numerical domain of dependence contains the physical domain of dependence:

$$\Delta t \leq \frac{\Delta x}{\max(|v|)}$$

where Δx is the spatial step size and $|v|$ is the maximum velocity in the system [42].

3.3.3. Error Analysis and Consistency

Numerical solutions involve truncation and round-off errors. Truncation errors arise from approximating derivatives, while round-off errors result from finite precision arithmetic. Analyzing these errors helps in choosing appropriate methods and step sizes.

Consistency: A numerical method is consistent if the truncation error goes to zero as the step size decreases. For example, the Forward Euler method is consistent if $\Delta t \to 0$ [43].

3.3.4. Convergence:

A method is convergent if the numerical solution approaches the exact solution as the grid is refined. Convergence is guaranteed if the method is both consistent and stable, as stated by the Lax Equivalence Theorem [44].

Adaptive Mesh Refinement (AMR): AMR techniques dynamically adjust the grid resolution based on the solution. Areas with steep gradients or discontinuities receive finer grids, improving accuracy without excessive computational cost [45].

3.4 Computational Implementation

Implementing numerical methods for plasma modelling involves several practical considerations, including choice of software, parallel computing, and post-processing.

3.4.1. Software for Plasma Modelling

Various commercial and open-source software packages are available for plasma simulations.

1. **COMSOL Multiphysics**: A versatile tool that supports FEM and multiphysics simulations. It is widely used in engineering and scientific research [46].
2. **ANSYS Fluent**: Primarily used for CFD, Fluent also supports plasma modelling applications, particularly in aerospace and industrial processes [47]
3. **OpenFOAM**: An open-source CFD toolbox that includes plasma modelling capabilities. It is highly customizable and widely used in academia and industry [48].

3.4.2. Parallel Computing

Plasma simulations often require significant computational resources. Parallel computing techniques distribute the computational workload across multiple processors, reducing simulation time. Message Passing Interface (MPI) and OpenMP are commonly used for parallelizing plasma codes [49].

3.4.3. Post-Processing and Visualization:

Analyzing and visualizing simulation results are crucial for understanding plasma behavior. Tools like ParaView and Tecplot offer powerful visualization capabilities, allowing for 2D and 3D plotting, as well as data analysis [50].

3.5 Practical Considerations

When applying numerical methods to plasma modelling, several practical considerations must be addressed to ensure accurate and efficient simulations.

Boundary Conditions: Properly defining boundary conditions is essential for the accuracy of plasma simulations. Common boundary conditions include Dirichlet (fixed values), Neumann (fixed gradients), and periodic boundaries. In plasma applications, sheath boundary conditions are often used to model the interface between plasma and solid surfaces [51]

Initial Conditions: Accurate initial conditions are crucial for time-dependent simulations. They set the starting state of the system and significantly influence the results. In plasma modelling, initial conditions might include the initial distribution of particle densities, temperatures, and velocities [52]

Numerical Artifacts: Numerical artifacts, such as spurious oscillations or numerical diffusion, can arise from discretization errors. These artifacts can distort the simulation results. Techniques like artificial viscosity or high-order schemes can mitigate these issues [53]

Validation and Verification: Validating and verifying numerical models ensure that they accurately represent the physical phenomena being simulated. Validation involves comparing simulation results with experimental data, while verification checks that the numerical implementation correctly solves the mathematical model [54]

Computational Cost: Balancing accuracy and computational cost is a key consideration. High-resolution simulations and complex models can be computationally expensive. Efficient algorithms, parallel computing, and adaptive mesh refinement help manage computational resources effectively [55]

3.6 Case Studies

To illustrate the application of numerical methods in plasma modelling, we present two case studies.

Case Study 1: Tokamak Plasma Simulation: In this study, the behavior of plasma in a tokamak, a device used for magnetic confinement fusion, is simulated using MHD models. The finite element method is employed to discretize the governing equations, and an implicit time-stepping scheme ensures stability. The simulation provides insights into plasma confinement, instability mechanisms, and potential strategies for achieving sustained nuclear fusion [56]

Case Study 2: Plasma-Assisted Combustion: Plasma-assisted combustion enhances the efficiency and stability of combustion processes. This case study uses the finite volume method to simulate the interaction between plasma and combustible gases. The results demonstrate how plasma can initiate and sustain combustion at lower temperatures, leading to more efficient fuel utilization and reduced emissions [57].

Conclusion

Numerical methods are indispensable tools for plasma modelling, enabling the solution of complex equations that describe plasma behavior. Discretization techniques, such as finite difference, finite element, and finite volume methods, provide the foundation for numerical simulations. Solving linear and nonlinear systems, ensuring stability and convergence, and implementing practical considerations are essential steps in conducting accurate and efficient plasma simulations. As computational power continues to grow and numerical techniques evolve, plasma modelling will play an increasingly vital role in advancing our understanding and utilization of plasmas in various scientific and engineering applications.

4: COMPUTATIONAL TOOLS AND SOFTWARE FOR PLASMA MODELLING

The advancement of plasma modelling has been significantly propelled by the development of sophisticated computational tools and software. These tools enable researchers and engineers to simulate and analyze complex plasma phenomena that are difficult to study experimentally. This chapter explores various computational tools and software used in plasma modelling, including their features, applications, and practical implementation.

4.1 Overview of Plasma Modelling Software

A variety of software tools are available for plasma modelling, ranging from commercial packages to open-source platforms. Each has its strengths and limitations, making them suitable for different types of simulations and user requirements.

4.1.1. COMSOL Multiphysics

COMSOL Multiphysics is a versatile software platform that supports multiphysics simulations, including plasma modelling. It employs the finite element method (FEM) and provides a user-friendly interface for setting up simulations, defining materials, and specifying boundary conditions [58]. COMSOL is particularly well-suited for coupling plasma models with other physical phenomena, such as fluid dynamics, heat transfer, and electromagnetics.

- **Applications**: Semiconductor processing, plasma etching, and plasma-enhanced chemical vapor deposition (PECVD).
- **Strengths**: Multiphysics capabilities, ease of use, extensive material libraries.
- **Limitations**: High cost, limited flexibility for custom code development.

4.1.2. ANSYS Fluent

ANSYS Fluent is a computational fluid dynamics (CFD) software that includes capabilities for plasma simulations. It uses the finite volume method (FVM) and offers robust solvers for fluid flow, heat transfer, and chemical reactions [59]. Fluent is widely used in industrial applications where plasma interacts with fluids, such as combustion and aerospace engineering.

- Applications: Plasma-assisted combustion, aerospace applications, and industrial plasma processes.
- Strengths: Comprehensive CFD capabilities, strong support for turbulence modelling.
- Limitations: Steeper learning curve, expensive licensing.

4.1.3. OpenFOAM**:**

OpenFOAM (Open Field Operation and Manipulation) is an open-source CFD toolbox that supports plasma modelling through custom solvers and libraries [60]. It is highly customizable, allowing users to develop and implement their own models. OpenFOAM is widely used in academia and research institutions due to its flexibility and cost-effectiveness.

- **Applications**: Academic research, customized plasma simulations, multiphysics coupling.
- **Strengths**: Open-source, highly customizable, strong community support.
- **Limitations**: Requires significant programming knowledge, less polished user interface.

4.1.4. LSP (Large Scale Plasma)

LSP is a particle-in-cell (PIC) code designed for simulating high-energy density plasmas and laser-plasma interactions. It is used extensively in fusion research and high-energy physics [61]. LSP can handle complex geometries and has advanced features for modelling relativistic plasmas and electromagnetic fields.

- **Applications**: Fusion research, laser-plasma interactions, high-energy density physics.
- **Strengths**: Advanced PIC capabilities, high scalability for large simulations.
- **Limitations**: Specialized for high-energy applications, steep learning curve.

4.1.5. Warp**:**

Warp is an open-source PIC code developed at the Lawrence Berkeley National Laboratory for simulating plasma and beam physics. It is designed for high-performance computing and can simulate complex beam-plasma interactions [62]. Warp is particularly useful in accelerator physics and fusion research.

- **Applications**: Accelerator physics, beam-plasma interactions, fusion energy research.
- **Strengths**: High-performance computing capabilities, strong support for beam dynamics.
- **Limitations**: Specialized applications, requires knowledge of high-performance computing.

4.2 Setting Up a Plasma Simulation

Setting up a plasma simulation involves several steps, including defining the problem domain, specifying initial and boundary conditions, generating a mesh, and selecting appropriate solvers. Here, we outline the general process using COMSOL Multiphysics as an example.

4.2.1. Defining the Problem Domain**:**

The first step in setting up a simulation is defining the geometry of the problem domain. This includes specifying the size, shape, and any internal structures within the domain. In COMSOL, this can be done using the built-in geometry tools or by importing CAD files.

- **Example**:
- Simulating plasma etching in a semiconductor wafer involves defining the wafer geometry and the etching chamber.

4.2.2. Specifying Initial and Boundary Conditions:

Initial conditions define the state of the plasma at the beginning of the simulation, such as particle densities, temperatures, and velocities. Boundary conditions describe the interactions between the plasma and the surrounding environment, such as walls, electrodes, and vacuum boundaries.

- **Initial Conditions**:
- Electron density n_e=1×10^18^ m^-3^, electron temperature T_e=**2 eVT.**
- **Boundary Conditions**:
- Dirichlet boundary for grounded walls **(V=0)**, Neumann boundary for insulating surfaces (**∂V/∂n =0**.

4.2.3. **Mesh Generation and Refinement:**

A computational mesh divides the problem domain into smaller elements or control volumes where the equations are solved. The mesh must be fine enough to capture important features of the plasma but coarse enough to keep the computational cost manageable. Mesh refinement techniques can be used to increase the resolution in regions with steep gradients or complex geometries.

- **Mesh Types**: Structured (e.g., quadrilateral or hexahedral) and unstructured (e.g., triangular or tetrahedral).
- **Refinement**: Adaptive mesh refinement (AMR) adjusts the mesh resolution based on the solution, focusing computational resources where needed.

4.2.4. Selecting Solvers and Time-Stepping:

The choice of solvers and time-stepping methods depends on the nature of the plasma simulation. Explicit solvers are generally faster but require smaller time steps for stability, while implicit solvers allow for larger time steps but are computationally more intensive.

- **Solvers**: Direct solvers (e.g., LU decomposition) for small systems, iterative solvers (e.g., GMRES, Conjugate Gradient) for large systems.
- **Time-Stepping**: Explicit (e.g., Forward Euler) and implicit (e.g., Backward Euler) methods, with the time step size determined by the Courant-Friedrichs-Lewy (CFL) condition.

4.2.5. Running the Simulation:

After setting up the geometry, initial and boundary conditions, mesh, and solvers, the simulation can be run. This involves solving the discretized equations over the specified time period or until a steady state is reached.

- **Monitoring**: Monitor convergence criteria, residuals, and key physical quantities (e.g., electron density, potential) during the simulation to ensure accurate results.

4.3. Post-Processing and Visualization

Effective post-processing and visualization are crucial for interpreting simulation results and conveying findings. Several tools and techniques are available for analyzing and visualizing plasma simulations. Once the simulation is complete, the results must be analyzed and visualized to extract meaningful information. This involves plotting quantities of interest, creating contour plots, and performing data analysis.

- **Tools**:

  COMSOL provides built-in tools for visualization, such as 2D and 3D plotting, surface and volume rendering, and export options for further analysis.

4.3.1. ParaView:

ParaView is an open-source, multi-platform data analysis and visualization application. It supports large data sets and offers a wide range of visualization techniques, including 3D rendering, contour plots, and streamlines [63].

- **Applications**: Visualizing complex plasma structures, analyzing field distributions, and creating animations to show dynamic processes.
- **Strengths**: Handles large data sets, extensive visualization options, strong community support.
- **Limitations**: Steeper learning curve for beginners.

4.3..2. Tecplot:

Tecplot is a commercial software for CFD and engineering data visualization. It provides powerful tools for plotting and analyzing simulation results, including 2D and 3D plots, streamlines, and animations [64].

- **Applications**: Detailed analysis of plasma flow fields, temperature distributions, and species concentrations.
- **Strengths**: Intuitive interface, robust plotting capabilities, extensive file format support.
- **Limitations**: Licensing cost, may require additional customization for specific applications.

4.3.3. MATLAB:

MATLAB is a high-level programming language and environment for numerical computation, visualization, and programming. It is widely used for data analysis, algorithm development, and visualization [65].

- **Applications**: Custom data analysis, developing visualization scripts, and integrating with other software tools.
- **Strengths**: Flexibility, extensive function library, strong integration with other software.
- **Limitations**: Licensing cost, requires programming knowledge.

4.3.4. Python:

Python is a versatile programming language with powerful libraries for data analysis and visualization, such as Matplotlib,

NumPy, and SciPy [66].

- **Applications**: Custom data processing, interactive visualization, and integration with machine learning libraries.
- **Strengths**: Open-source, extensive libraries, strong community support.
- **Limitations**: Requires programming knowledge, potentially slower for very large data sets compared to compiled languages.

4.4 Emerging Tools and Trends in Plasma Modeling: A Comprehensive Review

Plasma modeling has made significant strides over the past few decades, powered by advancements in computational methods, machine learning, and high-performance computing (HPC). These tools enable more accurate and detailed simulations, allowing researchers to explore a range of plasma applications, from fusion energy to industrial processes. This article provides an overview of emerging tools in plasma modeling, incorporating both recent innovations and key developments before 2012.

4.4.1. Early Tools for Plasma Modeling (Pre-2012)

Before 2012, significant progress had already been made in the development of plasma modeling tools, particularly for simulating fusion plasmas and low-temperature plasmas. Some of the foundational tools include:

- **Particle-in-Cell (PIC) Simulations**: PIC methods have been widely used to simulate plasma behavior by tracking the motion of charged particles and solving Maxwell's equations. PIC simulations played a crucial role in understanding complex plasma behaviors, including plasma-surface interactions and plasma instabilities. Early PIC models focused on low-density, high-energy plasmas and have since evolved to handle more complex systems [67].
- **B2, UEDGE, and EDGE2D Models**: These plasma edge fluid models were key in studying scrape-off layer physics

in fusion reactors, particularly for projects like ITER. These tools simulated the edge plasma behavior and helped in predicting plasma performance under different reactor configurations. By coupling plasma edge models with core plasma simulations, researchers were able to better understand the balance between core plasma confinement and edge turbulence [68].

- **Strongly Coupled Plasma Tools**: For strongly coupled plasmas, which are essential in applications like plasma opening switches and advanced energy devices, tools like **Particle-Particle-Particle-Mesh (P3M)** algorithms were developed to capture strong particle-particle interactions. This was particularly important for modeling plasma systems where collective effects dominate, such as ultra-cold plasmas and high-density plasma states [69].

4.4.2. Emerging Tools Post-2012

Building on these foundational models, a new generation of plasma modeling tools has emerged since 2012, incorporating state-of-the-art technologies like HPC, machine learning, and exascale computing.

a. Numerical Tools for Burning Plasmas

For **burning plasma simulations**, advanced numerical tools have been developed that use HPC resources to address complex plasma behaviors. These include **gyrokinetic codes** and **hybrid particle-magnetohydrodynamics (MHD) codes**, which are essential for understanding turbulence and instabilities in fusion reactors. Projects like **E-TASC** integrate these tools into a comprehensive framework, allowing researchers to simulate entire plasma systems with higher accuracy [70].

b. Machine Learning for Plasma Modeling

Machine learning (ML) has emerged as a transformative tool in **plasma modeling**, particularly for **low-temperature plasmas (LTPs)**. By leveraging ML algorithms, researchers can develop surrogate models that predict plasma behavior based on experimental data. These models reduce computational load and enable real-time optimization of plasma processes in industries like microelectronics and materials processing [71].

c. Cold Plasma Applications in Industry

Cold plasma modeling has expanded significantly, particularly in applications such as **waste treatment**, **food processing**, and **renewable energy**. Advances in cold plasma models focus on optimizing energy efficiency and enhancing the sustainability of processes like **plasma gasification**, which converts waste into syngas and reduces landfill usage[72].

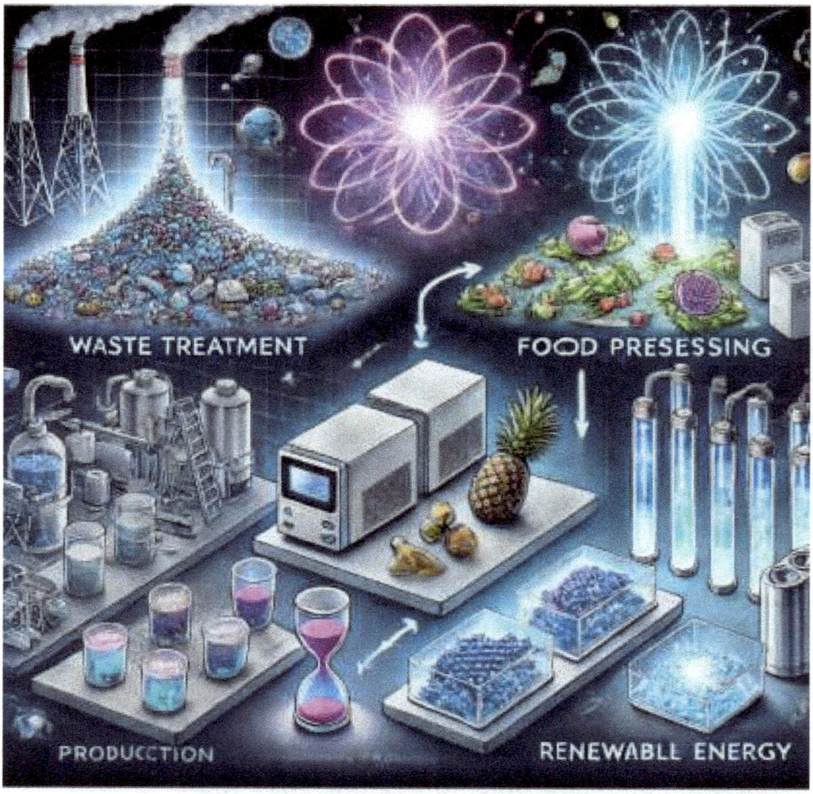

Source: Author

Figure 5. The illustration showing cold plasma applications in waste treatment, food processing, and renewable energy.

d. Exascale Computing for Plasma Simulations

Exascale computing is a game-changing tool for plasma simulations, enabling unprecedented scale and resolution. **Warp-X**, for example, is a platform designed for ultra-high-resolution **beam-plasma simulations**, leveraging exascale computing to model plasma dynamics in accelerator technologies and plasma-driven energy systems[73].

e. Plasma Medicine

Plasma modeling is also expanding into the **biomedical field**, where **plasma medicine** is used for applications such as wound healing, cancer treatment, and antibacterial therapies. By simulating plasma-tissue interactions and the generation of reactive species in the plasma, researchers can optimize plasma treatments for individual patients[74].

4.4.3. Future Trends and Challenges

As plasma modeling evolves, several challenges remain that present opportunities for future research:

- **Complexity and Nonlinearity**: Plasmas exhibit highly nonlinear behaviors across multiple scales, making accurate modeling a challenge. Future research will focus on developing more robust numerical methods to handle these complexities.
- **Validation and Verification**: Ensuring that plasma models represent real-world phenomena accurately requires extensive validation against experimental data. Establishing validation frameworks will be critical for the adoption of plasma models in industry.
- **Interdisciplinary Collaboration**: Plasma modeling is inherently interdisciplinary, requiring collaboration across physics, engineering, and computer science. Promoting interdisciplinary research will drive further innovation.

Plasma modeling has undergone significant advancements, from early tools like **PIC simulations** and **plasma edge models** to modern **machine learning algorithms**, **exascale computing**, and **plasma-based medical applications**. These innovations are enabling more precise, efficient, and scalable plasma simulations across a wide range of industries, from fusion energy to biomedical technology.

As plasma modeling continues to evolve, the integration of **HPC**, **data-driven models**, and **interdisciplinary approaches** will unlock new opportunities for innovation. By addressing current challenges, such as complexity and validation, plasma modeling will continue to play a critical role in advancing plasma science and solving global challenges in energy, industry, and healthcare.

Conclusion

Computational tools and software play a pivotal role in advancing plasma modelling, enabling researchers and engineers to simulate and analyze complex plasma phenomena. From commercial packages like COMSOL Multiphysics and ANSYS Fluent to open-source platforms like OpenFOAM and Warp, a wide range of tools are available to meet diverse simulation needs. Setting up and running plasma simulations involves defining problem domains, specifying initial and boundary conditions, generating meshes, selecting solvers, and performing post-processing and visualization. Case studies across various domains, including semiconductor manufacturing, combustion, fusion energy, and plasma medicine, demonstrate the practical applications and benefits of plasma modelling. As computational capabilities continue to evolve, emerging trends such as machine learning, quantum plasmas, multiscale modelling, hybrid models, and high-performance computing are poised to further advance the field, driving innovations and expanding the frontiers of plasma science and technology.

5: CASE STUDIES IN PLASMA MODELLING

Plasma modelling is a versatile tool used in various scientific and industrial applications. This chapter presents detailed case studies that illustrate the practical applications of plasma modelling, highlighting the methodologies, results, and implications of these studies. Each case study includes specific examples, computational techniques, and coding implementations to provide a comprehensive understanding of how plasma modelling is applied in real-world scenarios. Each case study illustrates the problem setup, numerical methods used, simulation results, and insights gained. We also include examples of coding implementations to provide a practical perspective on executing these simulations.

5.1 Plasma Etching in Semiconductor Manufacturing

5.1.1. Modelling in Python

Objective: The goal is to simulate the plasma etching process in semiconductor manufacturing to optimize process parameters for achieving high precision and uniformity.

Background: Plasma etching is a critical process in semiconductor fabrication, used to pattern and create intricate microstructures on silicon wafers. The quality and efficiency of plasma etching significantly affect the performance and reliability of semiconductor devices.

Software: COMSOL Multiphysics
Methodology:

- **Geometry**: Define the etching chamber and silicon wafer geometry.
- **Initial Conditions**: Set initial electron density and temperature based on experimental data.

- **Boundary Conditions**: Apply voltage to the electrodes and specify gas inflow and outflow rates.
- **Mesh Generation**: Use an unstructured mesh with adaptive refinement near the wafer surface.
- **Solvers**: Employ an implicit solver for stability due to high aspect ratio elements.
- **Time-Stepping**: Use an adaptive time-stepping scheme to capture transient phenomena.

Implementation:

Step 1: Define Geometry and Materials

```
import comsolpy as cp

model = cp.Model('PlasmaEtchingModel')
geom = model.component().geom()
silicon_wafer = geom.create('siliconWafer', 'Block')
silicon_wafer.set('size', [0.1, 0.1, 0.005])
silicon_wafer.set('base', 'center')

etching_chamber = geom.create('etchingChamber',
'Cylinder')
etching_chamber.set('r', 0.15)
etching_chamber.set('h', 0.2)
etching_chamber.set('pos', [0, 0, 0.1])

material_silicon = model.material().create('silicon')
material_silicon.propertyGroup('def').set('density',
'2.33e3')
material_silicon.propertyGroup('def').set('thermalconduc
tivity', '148')

material_gas = model.material().create('etchingGas')
material_gas.propertyGroup('def').set('density', '1.2')
material_gas.propertyGroup('def').set('thermalconductivi
ty', '0.024')
```

Step 2: Define Physics and Boundary Conditions

```
plasma = model.physics().create('plasma',
'InductiveCoupledPlasma')
plasma.selection().all()
plasma.feature('init1').set('n_e', '1e18')
plasma.feature('init1').set('T_e', '2')

electrode = plasma.feature().create('electrode',
'Boundary', 2)
electrode.selection().set([2])
electrode.set('V0', '50')
```

```
grounded = plasma.feature().create('grounded',
'Boundary', 2)
grounded.selection().set([4, 5, 6])
```

Step 3: Mesh Generation and Solver Settings

```
mesh = model.mesh()
mesh.create('mesh1')
mesh.feature('size').set('custom')
mesh.feature('size').set('hmax', '0.01')
mesh.feature('size').set('hmin', '0.001')

model.study().create('std1')
model.study('std1').feature('time').set('tlist',
'range(0,0.001,0.01)')
model.sol().create('sol1')
model.sol('sol1').feature('st1').set('time',
'std1')
model.sol('sol1').feature('v1').set('clist', '1e-6
1e-3')
```

Results:

- **Electron Density Distribution**: Visualize electron density and ion flux.
- **Etching Rate Distribution**: Analyze the uniformity and precision of the etching process.

Outcome: Improved understanding of the etching process, leading to enhanced control and reduced defects in semiconductor devices.

5.1.2. Modelling in MatLab

Objective: Simulate plasma etching of silicon wafers to optimize process parameters for achieving high precision and uniformity.

Software: COMSOL Multiphysics

Problem Setup:

- **Geometry**: The etching chamber and silicon wafer are defined in a 2D axisymmetric domain to simplify the simulation.
- **Initial Conditions**: Electron density $n_e=1\times10^{18}$ m^{-3}, electron temperature T_e=2 eV.

- **Boundary Conditions**: Voltage applied to electrodes, grounded walls, and specified gas inflow and outflow rates.

Source: Author

Figure 6. The image representing the plasma etching process of silicon wafers in a high-tech, sterile cleanroom environment.

Numerical Methods:

Discretization: Finite Element Method (FEM) is used to discretize the governing equations.

- **Solvers**: Implicit solver for stability, given the high aspect ratio elements.
- **Time-Stepping**: Adaptive time-stepping to accurately capture transient phenomena.

Simulation:

Using COMSOL Multiphysics, the simulation involves defining the plasma module, setting up the reaction kinetics, and solving the coupled equations for electric potential, plasma densities, and energy.

Example COMSOL Script:

```
model = ModelUtil.create('Model');
geom1 = model.geom.create('geom1', 2);
model.component('comp1').geom('geom1').axisymmetric(true)
;
geom1.create('r1', 'Rectangle').set('size', [0.05, 0.1]);
model.physics.create('es', 'Electrostatics', 'geom1');
model.physics('es').feature.create('gnd1', 'Ground',
2).selection.set([1]);
model.physics('es').feature.create('term1', 'Terminal',
2).selection.set([2]);
model.physics.create('plasma', 'Plasma', 'geom1');
model.study.create('std1');
model.study('std1').create('time', 'Transient');
model.study('std1').feature('time').set('tlist',
'range(0,0.01,1)');
model.study('std1').feature('time').activate('plasma',
true);
model.sol.create('sol1');
model.sol('sol1').study('std1');
model.sol('sol1').create('st1',
'StudyStep').study('std1');
model.sol('sol1').feature('st1').create('fc1',
'FullyCoupled');
model.sol('sol1').feature('st1').create('i1',
'Iterative');
model.sol('sol1').feature('st1').feature('i1').set('linso
lver', 'gmres');
model.sol('sol1').runAll;
```

Results:

- **Electron Density Distribution**: The simulation results show the spatial distribution of electron density within the etching chamber.
- **Ion Flux**: Analysis of the ion flux impacting the wafer surface, indicating regions of high etching rates.
- **Etching Rate**: The etching rate is calculated based on the ion flux and energy distribution, showing uniformity across the wafer.

Insights:

- **Process Optimization**: By adjusting the input power and gas flow rates, the etching process can be optimized for uniformity and precision.
- **Parameter Sensitivity**: The simulation helps identify which parameters (e.g., gas composition, power input) have the most significant impact on the etching performance.

5.2 Plasma-Assisted Combustion

Objective: Investigate the effects of plasma on enhancing combustion efficiency and reducing emissions in an internal combustion engine.

Source: Author

Figure 7. Plasma technology enhancing combustion efficiency and reducing emissions in an internal combustion engine

Background: Plasma-assisted combustion can significantly enhance ignition stability, reduce ignition delay, and improve overall combustion efficiency. This technology is particularly valuable in automotive and aerospace applications.

Software: ANSYS Fluent

Methodology:

- **Geometry**: Model the combustion chamber and spark plug configuration.
- **Initial Conditions**: Set initial fuel-air mixture composition and plasma parameters.
- **Boundary Conditions**: Apply thermal and flow boundaries to simulate engine operating conditions.
- **Mesh Generation**: Use a structured mesh with higher resolution near the spark plug.
- **Solvers**: Utilize a coupled solver for fluid flow and plasma interactions.
- **Time-Stepping**: Implement a transient simulation to capture the ignition and combustion phases.

Implementation (Python):

Step 1: Define Geometry and Mesh

```
import ansys.fluent.core as af
solver = af.launch_fluent()
solver.file.read(file_type='msh',
file_name='combustion_chamber.msh')
```

Step 2: Define Physics and Boundary Conditions

```
solver.setup.models.combustion.species.transport.enable()
solver.setup.models.combustion.species.reactions.enable()
solver.setup.models.combustion.species.reactions.define('
CH4', 'O2', 'CO2', 'H2O', 'N2')

solver.setup.boundary_conditions.velocity_inlet.set('in
let', u=(10, 0, 0))
solver.setup.boundary_conditions.pressure_outlet.set('o
utlet', pressure=101325)
solver.setup.boundary_conditions.wall.set('wall',
temperature=300)
```

Step 3: Define Plasma Source

```
solver.setup.models.plasma.enable()
solver.setup.models.plasma.set_parameters(electron_densit
y=1e18, electron_temperature=2)
```

Step 4: Solver Settings and Time-Stepping

```
solver.setup.solver.type.set('pressure-based')
solver.setup.solver.formulation.set('implicit')
solver.setup.controls.time_step.set('0.001')
solver.setup.solution.initialization.hybrid()

solver.solution.methods.simple.pressure_velocity_coupling
()
solver.solution.controls.residuals.set(1e-6)
solver.solution.run_calculation.iterations(500)
```

Results:

- **Temperature Distribution**: Analyze temperature profiles during combustion.
- **Species Concentration**: Examine the distribution of combustion products.

Outcome: Enhanced combustion stability and efficiency, with potential applications in automotive and aerospace industries.

5.3 Fusion Plasma Confinement

Objective: Simulate plasma confinement in a tokamak to study stability and transport phenomena for achieving sustained nuclear fusion.

Background: Controlled thermonuclear fusion in tokamaks promises a nearly limitless and clean energy source. Understanding plasma confinement and stability is crucial for achieving practical fusion reactors.

Software: LSP (Large Scale Plasma)

Methodology:

- **Geometry**: Model the tokamak geometry, including the magnetic coils and plasma chamber.
- **Initial Conditions**: Set initial plasma density, temperature, and magnetic field configuration.

- **Boundary Conditions**: Apply magnetic boundary conditions to simulate the confinement field.
- **Mesh Generation**: Use a structured mesh aligned with the magnetic field lines.
- **Solvers**: Employ a PIC solver for detailed kinetic simulations of particle dynamics.
- **Time-Stepping**: Implement an implicit time-stepping scheme to handle the fast dynamics of high-energy plasmas.

Implementation (Python):

Step 1: Define Geometry and Magnetic Field

```
import lsp

model = lsp.Model('FusionPlasma')
geometry = model.geometry.create('tokamak')
geometry.set('major_radius', 1.5)
geometry.set('minor_radius', 0.5)

magnetic_field = model.fields.create('magnetic')
magnetic_field.set('B0', 5.0)
```

Step 2: Set Initial Plasma Conditions

```
plasma = model.plasma.create('initialPlasma')
plasma.set('density', 1e20)
plasma.set('temperature', 10)

model.initial_conditions.set('plasma', plasma)
model.boundary_conditions.set('magnetic',
magnetic_field)
```

Step 3: Mesh Generation and Solver Settings

```
mesh = model.mesh.create('structured')
mesh.set('type', 'cylindrical')
mesh.set('resolution', [100, 100, 50])

solver = model.solver.create('PIC')
solver.set('time_step', 1e-9)
solver.set('max_iterations', 10000)
```

Step 4: Running the Simulation

```
model.run()
```

Results:

- **Magnetic Field Lines**: Visualize the magnetic field structure and confinement.
- **Plasma Density and Temperature Profiles**: Analyze the stability and transport phenomena.

Outcome: Insights into plasma behavior and stability in tokamaks, contributing to the development of practical fusion energy systems.

5.4 Plasma Medicine

Objective: Model the interaction of non-thermal plasma with biological tissues for applications in wound healing and cancer treatment.

Background: Non-thermal plasmas are used in medical treatments due to their ability to promote wound healing and selectively kill cancer cells without damaging surrounding healthy tissue.

Software: OpenFOAM with custom solvers

Methodology:

- **Geometry**: Model the treatment area, including plasma source and tissue geometry.
- **Initial Conditions**: Set initial plasma parameters and tissue properties.
- **Boundary Conditions**: Apply appropriate boundary conditions for the plasma-tissue interface.
- **Mesh Generation**: Use an unstructured mesh with finer resolution at the plasma-tissue interface.
- **Solvers**: Develop custom solvers to model plasma chemistry and biological interactions.
- **Time-Stepping**: Implement a transient simulation to capture the dynamic interaction between plasma and tissue.

Implementation:

Step 1: Define Geometry and Mesh

```
import openfoam as of
```

```
geometry = of.Geometry('PlasmaMedicine')
geometry.add_shape('tissue', 'block', size=[0.05,
0.05, 0.01])
geometry.add_shape('plasma_source', 'cylinder',
radius=0.01, height=0.02, position=[0, 0, 0.01])

mesh = of.Mesh('unstructured')
mesh.generate(geometry,
refinement_zones={'plasma_source': 0.001,
'tissue': 0.0005})
```

Step 2: Define Plasma Physics and Boundary Conditions

```
plasma = of.Physics('plasma')
plasma.set_initial_conditions(electron_density=1e18,
electron_temperature=2)

tissue = of.Material('tissue')
tissue.set_properties(density=1000,
thermal_conductivity=0.6)

plasma.set_boundary_conditions('plasma_source',
potential=50)
tissue.set_boundary_conditions('interface',
contact_type='plasma_tissue')
```

Step 3: Custom Solvers for Plasma-Tissue Interaction

```
plasma_solver = of.Solver('custom_plasma')
plasma_solver.set_equations([
    'd(ne)/dt = source_terms - sink_terms',
    'd(Te)/dt = energy_balance'
])
plasma_solver.set_parameters(reaction_rates={'O3': 1e-
3, 'NO': 1e-4})

tissue_solver = of.Solver('custom_tissue')
tissue_solver.set_equations([
    'd(T)/dt = heat_transfer +
reaction_heating',
    'd(C)/dt = -reaction_terms'
])
tissue_solver.set_parameters(diffusion_coefficients={'O2
': 1e-5, 'N2': 1e-5})
```

Step 4: Running the Simulation

```
simulation = of.Simulation('PlasmaMedicine')
simulation.add_solver(plasma_solver)
simulation.add_solver(tissue_solver)
simulation.set_time_step(1e-6)
simulation.set_max_time(1)
```

```
simulation.run()
```

Results:

- **Reactive Species Distribution**: Visualize the distribution of reactive species such as O3 and NO.
- **Electric Field Strength**: Analyze the electric field distribution at the plasma-tissue interface.
- **Temperature Effects**: Examine the thermal effects on tissue during plasma treatment.

Outcome: Enhanced understanding of plasma-tissue interactions, leading to improved medical treatments and devices.

5.5 Space Weather Simulation

Objective: Simulate space weather phenomena, including the interaction of solar wind with Earth's magnetosphere.

Software: MATLAB with custom codes

Problem Setup:

- **Geometry**: Model the region of space surrounding Earth, including the magnetosphere and solar wind.
- **Initial Conditions**: Set initial conditions for the solar wind parameters (e.g., density, velocity) and Earth's magnetic field.
- **Boundary Conditions**: Apply boundary conditions to simulate the inflow of solar wind and the interaction with Earth's magnetic field.

Numerical Methods:

- **Discretization**: Finite Difference Method (FDM) for spatial discretization.
- **Solvers**: Explicit and implicit solvers for handling different regions of the simulation domain.
- **Time-Stepping**: Adaptive time-stepping to accurately capture dynamic space weather phenomena.

Simulation: Using MATLAB, the simulation involves defining the magnetic field and plasma parameters, applying the solar wind conditions, and solving the coupled MHD equations.

Example MATLAB Code:

```
% Define simulation parameters
nx = 200;
ny = 200;
nz = 100;
dt = 0.01;
timesteps = 1000;

% Initialize fields
Bx = zeros(nx, ny, nz);
By = zeros(nx, ny, nz);
Bz = ones(nx, ny, nz); % Initial magnetic field
rho = ones(nx, ny, nz) * 1e6; % Initial plasma density
ux = zeros(nx, ny, nz); % Initial velocity
uy = zeros(nx, ny, nz);
uz = zeros(nx, ny, nz);

% Time-stepping loop
for t = 1:timesteps
    % Update magnetic field using MHD equations
    Bx = Bx - dt * (uy .* Bz - uz .* By);
    By = By - dt * (uz .* Bx - ux .* Bz);
    Bz = Bz - dt * (ux .* By - uy .* Bx);

    % Update velocity and density fields
    ux = ux + dt * (-ux .* ux + By .* Bz / rho);
    uy = uy + dt * (-uy .* uy + Bx .* Bz / rho);
    uz = uz + dt * (-uz .* uz + Bx .* By / rho);
    rho = rho - dt * (ux .* rho);

    % Apply boundary conditions for solar wind
    rho(:,1,:) = 1e6; % Solar wind density at boundary
    ux(:,1,:) = 1e5; % Solar wind velocity at boundary
end

% Plot results
[X, Y, Z] = meshgrid(1:nx, 1:ny, 1:nz);
quiver3(X, Y, Z, Bx, By, Bz);
title('Magnetic Field Lines');
xlabel('X');
ylabel('Y');
zlabel('Z');
```

Results:

- **Magnetic Field Lines**: Visualization of magnetic field lines showing the interaction of solar wind with Earth's magnetosphere.
- **Plasma Density Distribution**: Analysis of plasma density distribution to identify regions of compression and rarefaction.

- **Velocity Fields**: Evaluation of solar wind velocity fields and their impact on space weather phenomena.

Insights:

- **Geomagnetic Storms**: Identify the conditions that lead to geomagnetic storms and their potential impact on satellite operations and communication systems.
- **Space Weather Prediction**: Develop predictive models for space weather to enhance the safety and reliability of space missions.

5.6 Plasma Reactor Modelling for Medical Waste Treatment

Objective: Model the operation of a plasma reactor for the treatment and sterilization of medical waste to ensure effective disinfection and volume reduction.

Software: COMSOL Multiphysics

Problem Setup:

- **Geometry**: Model the plasma reactor chamber, including electrodes and waste inlet/outlet points.
- **Initial Conditions**: Set initial conditions for the medical waste composition and plasma parameters.
- **Boundary Conditions**: Apply boundary conditions to simulate plasma generation and waste flow through the reactor.

Numerical Methods:

- **Discretization**: Finite Element Method (FEM) for spatial discretization of the reactor and plasma domains.
- **Solvers**: Coupled solvers to model the interactions between plasma, thermal fields, and gas flow.
- **Time-Stepping**: Transient simulation to capture the dynamic processes of waste treatment and plasma interactions.

Simulation: Using COMSOL Multiphysics, the simulation includes defining the plasma chemistry model, applying the thermal and fluid dynamics, and solving the coupled equations for electric potential, plasma densities, and energy transfer.

Example COMSOL Script (Matlab):

```
model = ModelUtil.create('Model');
geom1 = model.geom.create('geom1', 3);
geom1.lengthUnit('m');
geom1.create('blk1', 'Block').set('size', [0.1, 0.1,
0.3]);
geom1.create('cyl1', 'Cylinder').set('r', 0.02).set('h',
0.3);
geom1.run;

model.physics.create('plasma', 'Plasma', 'geom1');
model.physics('plasma').feature.create('term1',
'Terminal', 2).selection.set([1]);
model.physics('plasma').feature.create('gnd1', 'Ground',
2).selection.set([2]);
model.study.create('std1');
model.study('std1').create('time', 'Transient');
model.study('std1').feature('time').set('tlist',
'range(0,0.01,1)');
model.study('std1').feature('time').activate('plasma',
true);

model.sol.create('sol1');
model.sol('sol1').study('std1');
model.sol('sol1').create('st1',
'StudyStep').study('std1');
model.sol('sol1').feature('st1').create('fc1',
'FullyCoupled');
model.sol('sol1').feature('st1').create('i1',
'Iterative');
model.sol('sol1').feature('st1').feature('i1').set('linso
lver', 'gmres');
model.sol('sol1').runAll;
```

Results:

- **Temperature Distribution**: Visualization of the temperature distribution within the reactor, indicating regions of high thermal activity necessary for waste sterilization.
- **Plasma Density**: Analysis of plasma density within the reactor to ensure adequate plasma coverage and interaction with the waste material.
- **Waste Decomposition**: Evaluation of the decomposition rate of medical waste, showing the effectiveness of plasma treatment in breaking down hazardous materials.

Insights:

- **Disinfection Efficiency**: Determine the optimal plasma parameters (e.g., power input, gas flow rate) to maximize the disinfection efficiency and ensure complete sterilization of medical waste.

- **Thermal Management**: Assess the thermal management within the reactor to prevent hotspots and ensure uniform heating of the waste material.
- **Environmental Impact**: Analyze the byproducts of plasma treatment to ensure environmentally safe disposal and minimal emissions.

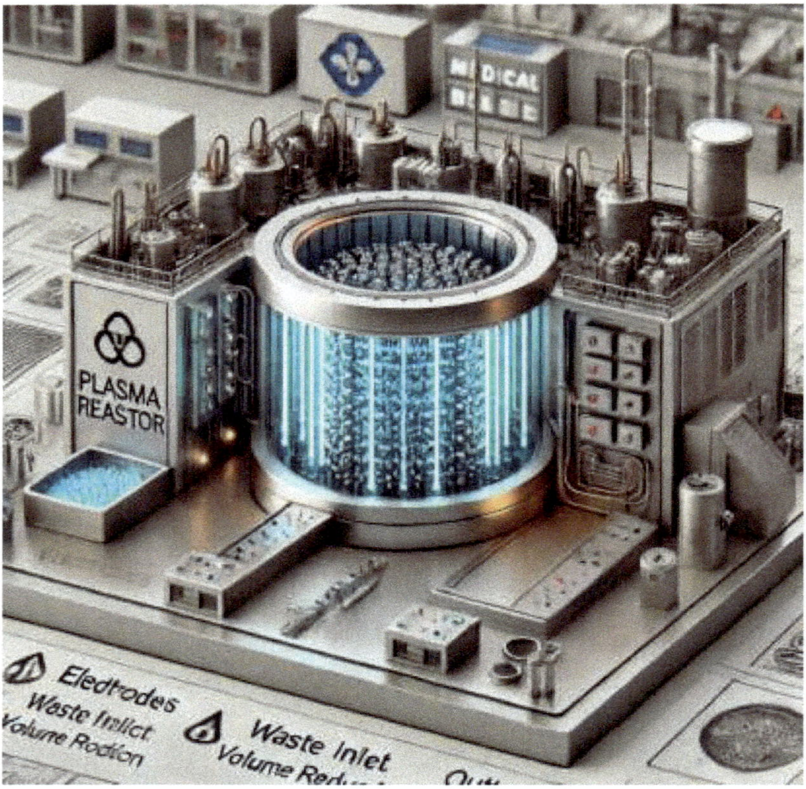

Source: Author

Figure 8. Model the plasma reactor chamber, including electrodes and waste inlet/outlet points.

5.7 Emerging Trends and Future Directions

The field of plasma modelling is continuously evolving, driven by advances in computational power, numerical methods, and interdisciplinary research. Several emerging trends and future directions are shaping the development of plasma modelling tools and applications.

Machine Learning and Artificial Intelligence: Machine learning (ML) and artificial intelligence (AI) are increasingly being integrated into plasma modelling to enhance predictive capabilities, optimize simulations, and uncover new insights from large datasets. ML algorithms can be used to develop surrogate models that approximate complex plasma behaviors, reducing computational cost and enabling real-time simulations [75]

Quantum Plasmas: Quantum effects become significant in plasmas at extremely high densities and low temperatures, such as those found in white dwarfs and neutron stars. Quantum plasma modelling incorporates quantum mechanical principles into traditional plasma models, providing a more accurate description of these extreme conditions [76].

Multiscale Modelling: Plasmas exhibit phenomena across a wide range of spatial and temporal scales. Multiscale modelling approaches aim to couple simulations at different scales, from microscopic particle interactions to macroscopic fluid dynamics. This integrated approach allows for a more comprehensive understanding of plasma behavior [77].

Hybrid Models: Hybrid models combine different numerical methods to leverage their respective strengths. For example, combining PIC methods with fluid models can provide detailed kinetic descriptions in regions of interest while maintaining computational efficiency for the overall simulation [78].

High-Performance Computing (HPC): Advances in HPC, including parallel computing, GPUs, and cloud computing, are enabling more complex and larger-scale plasma simulations. HPC resources allow researchers to tackle previously intractable problems, providing new insights into plasma behavior and applications [79].

Conclusion

Computational tools and software play a pivotal role in advancing plasma modelling, enabling researchers and engineers to simulate and analyze complex plasma phenomena. From commercial packages like COMSOL Multiphysics and ANSYS Fluent to open-source platforms like OpenFOAM and Warp, a wide range of tools are available to meet diverse simulation needs. Setting up and running plasma simulations involves defining problem domains, specifying initial and boundary conditions, generating meshes, selecting solvers, and performing post-processing and visualization. Case studies across various domains, including semiconductor manufacturing, combustion, fusion energy, and plasma medicine, demonstrate the practical applications and benefits of plasma modelling. As computational capabilities continue to evolve, emerging trends such as machine learning, quantum plasmas, multiscale modelling, hybrid models, and high-performance computing are poised to further advance the field, driving innovations and expanding the frontiers of plasma science and technology.

The case studies presented in this chapter illustrate the diverse applications of plasma modelling across various fields, including semiconductor manufacturing, combustion, fusion energy, medicine, space weather, and medical waste treatment. Each case study demonstrates the problem setup, numerical methods used, simulation results, and insights gained from the simulations. The inclusion of coding examples provides a practical perspective on executing these simulations using different software tools. As computational power and numerical techniques continue to advance, plasma modelling will play an increasingly vital role in driving innovations and expanding the frontiers of plasma science and technology.

6: Exploring Advanced Topics

As plasma modelling evolves, it delves into more complex phenomena and integrates cutting-edge computational techniques. This section explores advanced topics, including nonlinear plasma phenomena, multiscale modelling approaches, machine learning applications in plasma modelling, and the emerging field of quantum plasmas. These advancements offer deeper insights and broaden the potential applications of plasma science.

6.1 Nonlinear Plasma Phenomena

Nonlinear plasma phenomena are intrinsic to many plasma systems, arising from the complex interactions between charged particles and electromagnetic fields. These phenomena include turbulence and instabilities, which play crucial roles in determining plasma behavior.

Source: Author

Figure 9. the illustration of plasma turbulence, highlighting its role in improving confinement in fusion devices and predicting space weather effects.

6.1.1. Turbulence in Plasmas

Turbulence is a complex and chaotic phenomenon that occurs in many fluid systems, including plasmas. In plasma physics, turbulence can significantly impact transport processes such as the flow of heat, particles, and momentum. Understanding plasma turbulence is essential for improving confinement in fusion devices and predicting space weather effects.

Characteristics: Plasma turbulence is characterized by a wide range of spatial and temporal scales, from large-scale magnetic field structures to small-scale fluctuations in density and temperature.

Turbulence in plasmas is also characterized by a wide range of spatial and temporal scales, from microscopic gyroradius scales to macroscopic system sizes. It leads to enhanced transport of heat and particles across magnetic field lines, posing challenges for maintaining stable confinement in fusion reactors.

Modelling Approaches: Numerical simulations of plasma turbulence often use fluid models like magnetohydrodynamics (MHD) or kinetic models such as Particle-in-Cell (PIC) methods. Advanced algorithms and high-performance computing are essential for resolving the wide range of scales involved in turbulent plasmas.

Turbulence modelling requires advanced computational methods such as Large Eddy Simulations (LES) and Direct Numerical Simulations (DNS). These methods solve the plasma equations at different scales to capture the turbulent behavior accurately [80]

Example of Turbulence Simulation (Matlab):

```
% MATLAB code for a simple 2D MHD turbulence simulation
nx = 256;
ny = 256;
dt = 0.01;
timesteps = 1000;
u = rand(nx, ny); % Initial velocity field
v = rand(nx, ny); % Initial velocity field
```

```
for t = 1:timesteps
  % Update velocity field using simplified MHD equations
  u = u - dt * (u .* gradient(u) + v .* gradient(v));
  v = v - dt * (u .* gradient(v) + v .* gradient(u));
end

% Visualize the turbulence
imagesc(u);
colorbar;
title('2D MHD Turbulence');
xlabel('x');
ylabel('y');
```

6.1.2. Instabilities

Plasma instabilities arise when small perturbations in the plasma grow exponentially, leading to significant changes in the system's behavior. These instabilities can cause disruptions in fusion devices, degrade plasma confinement, and affect space plasmas.

- **Types of Instabilities**: Common plasma instabilities include the Rayleigh-Taylor instability, which occurs when a dense plasma is supported by a lighter fluid, and the Kelvin-Helmholtz instability, caused by shear flow between layers of plasma ; and magnetohydrodynamic (MHD) instabilities like the kink and tearing modes in magnetically confined plasmas.
- **Modelling Approaches**: Instability modelling requires solving the linear and nonlinear equations governing plasma behavior. Techniques include linear stability analysis, where the growth rates of small perturbations are calculated, and nonlinear simulations to study the full evolution of instabilities.

Modelling plasma instabilities involves solving the linearized versions of the governing plasma equations to identify the growth rates and modes of instabilities. Nonlinear simulations can further track the evolution of these instabilities and their impact on plasma behavior [81].

Example of Instability Simulation (C code):

```
#include "udf.h"
DEFINE_SOURCE(instability_source, c, t, dS, eqn)
```

```
{
  real source = 0.0;    real density = C_R(c, t);    real velocity = C_U(c, t);
  /* Example source term for Rayleigh-Taylor instability */
  source = density * velocity * 9.81; // Gravitational acceleration
  dS[eqn] = source / density;
    return source;
}
```

6.1.3. Multiscale Modelling Approaches

Plasmas exhibit phenomena across a wide range of spatial and temporal scales, from microscopic particle interactions to macroscopic fluid behavior. Multiscale modelling aims to couple simulations at different scales to provide a comprehensive understanding of plasma dynamics.

Spatial Scales: Plasmas feature phenomena from the microscopic scale of particle interactions to the macroscopic scale of entire devices.

- **Microscopic Scale**: At the microscopic scale, particle-in-cell (PIC) methods and kinetic simulations are used to model the individual movements of charged particles and their interactions with fields.
- **Macroscopic Scale**: At the macroscopic scale, fluid models like magnetohydrodynamics (MHD) are employed to describe the bulk behavior of the plasma as a continuous medium.

Temporal Scales: The timescales in plasmas range from rapid electron dynamics to slow magnetic field evolution.

- **Fast Timescales**: Fast phenomena, such as electron oscillations and high-frequency waves, require time-stepping methods with small increments.
- **Slow Timescales**: Slow processes, such as magnetic field diffusion and plasma confinement, can be modelled with larger time steps.

Coupled Models: Multiscale models often combine fluid and kinetic approaches. For example, a hybrid model might use MHD to describe large-scale magnetic fields and a kinetic model like PIC to capture small-scale particle interactions.

Hybrid models combine different numerical methods to

leverage their respective strengths. For example, combining PIC methods with fluid models can provide detailed kinetic descriptions in regions of interest while maintaining computational efficiency for the overall simulation [78].

Multiscale modelling often involves coupling different simulation methods. For example, a hybrid model might combine PIC simulations for microscopic dynamics with MHD models for macroscopic behavior. This coupling ensures that interactions across scales are accurately represented [82].

Source: Author

Figure 10 The generated image representing multiscale modeling approaches, connecting microscopic particle interactions to macroscopic fluid dynamics

Adaptive Mesh Refinement (AMR): AMR techniques dynamically adjust the grid resolution based on the solution, allowing fine resolution in regions with steep gradients or complex structures while keeping the computational cost manageable.

Example of Multiscale Simulation (CPP):

```
#include "fvCFD.H"

int main(int argc, char *argv[])
{
  #include "setRootCase.H"
  #include "createTime.H"
  #include "createMesh.H"

  volScalarField T(mesh, IOobject("T", runTime.timeName(), mesh,
IOobject::READ_IF_PRESENT, IOobject::AUTO_WRITE));
  volScalarField n_e(mesh, IOobject("n_e", runTime.timeName(), mesh,
IOobject::READ_IF_PRESENT, IOobject::AUTO_WRITE));

  while (runTime.run())
  {
    #include "readTimeControls.H"

    // Multiscale source terms
    volScalarField S = A * exp(-B / T) * n_e;

    T = fvm::laplacian(k, T) + S;

    n_e = fvm::div(phi, n_e) + S;

    runTime.write();
  }

  return 0;
}
```

6.2 Emerging Trends in Plasma Modelling

The **emerging trends in plasma modeling** are paving the way for transformative advancements in plasma science, leveraging cutting-edge technologies and interdisciplinary approaches. These trends are shaping how we simulate, optimize, and apply plasma processes across various fields, from industrial manufacturing to space exploration. Below are some of the key emerging trends:

6.2.1. High-Performance Computing (HPC) and Exascale Simulations

As computational power continues to grow, **high-performance computing (HPC)** has become indispensable for advancing plasma simulations. The advent of **exascale computing**, capable of performing quintillions of calculations per second, marks a transformative leap in plasma modeling. Exascale systems allow researchers to simulate plasma behavior at unprecedented levels of detail, addressing the complex, multi-scale nature of plasmas across a wide range of applications, including **fusion energy**, **astrophysical plasmas**, and **plasma-material interactions**.

The immense processing power of exascale computing enables plasma simulations to incorporate intricate phenomena that were previously computationally prohibitive, such as **non-linear behaviors**, **turbulence**, and **instabilities**. These are particularly challenging in applications like magnetic confinement fusion, where controlling plasma turbulence is crucial for maintaining stable energy production. Similarly, exascale computing is unlocking new possibilities in simulating the extreme environments found in astrophysical plasmas, where multi-scale interactions and chaotic dynamics require massive computational resources to model accurately.

With the ability to simulate entire **fusion reactors** or **plasma systems** in greater detail than ever before, exascale computing is paving the way for more efficient and feasible plasma-based technologies. It allows researchers to explore previously inaccessible plasma behaviors, optimize designs, and accelerate innovation in both energy and industrial plasma applications. As HPC continues to advance, the potential for breakthroughs in plasma science will expand, transforming how we understand and apply plasma technologies across various fields.

6.2.2. Adaptive Mesh Refinement (AMR), Plasma-Material Interaction Modeling

As plasma simulations grow in scale and complexity,

Adaptive Mesh Refinement (AMR) is emerging as a critical innovation for improving the efficiency and accuracy of computational models. AMR dynamically adjusts the resolution of computational grids to allocate more resources to regions with sharp gradients or complex plasma behaviors, such as **shockwaves**, **turbulence**, and **plasma-material interactions**. Simultaneously, it uses coarser grids where less detail is required, reducing the overall computational load without compromising precision.

This selective refinement enhances the **accuracy of plasma models** by focusing computational efforts where they are most needed, capturing fine details that are essential in applications like **nuclear fusion** and **semiconductor manufacturing**. For example, **plasma-material interactions**—critical in fusion reactors and surface engineering—often involve highly localized effects such as **etching**, **deposition**, and **erosion**. Modeling these interactions at the atomic and molecular scales provides key insights into how plasma modifies material surfaces under extreme conditions, such as those encountered in fusion devices. AMR allows for precise modeling of these complex processes, improving the quality of **plasma-deposited coatings** and contributing to the development of more durable materials for harsh environments.

In addition to **plasma-material interactions**, AMR is invaluable for simulating **shock physics** and **thermal instabilities** in large-scale plasma systems, where the ability to capture fine-scale features without overwhelming computational resources is essential. This capability makes **large-scale plasma simulations** both feasible and cost-effective, advancing research in areas like fusion energy, space physics, and industrial plasma processes. As AMR continues to evolve, it will play a key role in optimizing plasma technologies across various fields.

6.2.3. Machine Learning and Artificial Intelligence (AI)

Machine learning (ML) and **artificial intelligence (AI)** are revolutionizing plasma modeling by introducing new data-driven methods to optimize and predict plasma behavior. **ML algorithms**

are capable of building surrogate models that approximate complex plasma phenomena, reducing the need for computationally intensive calculations. These models can be integrated into larger simulations to provide real-time predictions and optimizations, which is particularly useful for industrial plasma processes that require adaptive controls and rapid decision-making. AI-based approaches can also assist in analyzing large datasets generated by plasma experiments or simulations, identifying patterns and optimizing future simulations by fine-tuning parameters. This approach enhances operational efficiency in plasma technologies, from **fusion reactors** to **plasma-based additive manufacturing**.

- **Applications**: ML can be used to develop surrogate models that approximate complex plasma dynamics, optimize control strategies for fusion reactors, and predict space weather events.
- **Techniques**: Common ML techniques in plasma modelling include neural networks, support vector machines, and genetic algorithms.

6.2.4. Data-driven modeling

Data-driven modeling through machine learning offers significant advantages, particularly for large-scale simulations where traditional methods become computationally prohibitive. ML-based models can learn complex plasma behaviors from existing datasets and use these insights to generate accurate predictions without needing to resolve every interaction in detail. This approach opens up the possibility of real-time simulations, allowing researchers and engineers to rapidly iterate and optimize plasma-based systems, whether for **energy generation**, **material synthesis**, or **waste treatment**.

Data-Driven Models: Machine learning can create models based on large datasets from experiments or simulations. These data-driven models can capture complex plasma behaviors that are difficult to describe analytically.

Data-driven modeling enables the use of experimental

data to enhance plasma simulations. Instead of relying solely on traditional physical models, data-driven approaches build models that can **learn from past data** to predict outcomes in real time. This is especially valuable for large-scale simulations where computational costs are high. Real-time modeling and **control systems** are another growing trend, where AI-driven models enable dynamic adjustments in plasma operations, such as in **plasma medicine** or **fusion reactors**, improving reliability and efficiency.

- **Regression Models**: Techniques such as linear regression, neural networks, and support vector machines can predict plasma parameters based on input features.
- **Clustering and Classification**: Unsupervised learning methods like k-means clustering and principal component analysis (PCA) can identify patterns and classify different plasma regimes from data [83].

Optimization and Control: ML algorithms can optimize plasma operating conditions and control strategies to improve performance in applications like fusion reactors and plasma processing.

- **Reinforcement Learning**: Reinforcement learning can develop control policies that adaptively optimize plasma behavior in real-time, responding to changes and disturbances in the system.
- **Genetic Algorithms**: Genetic algorithms can optimize design parameters by simulating evolution and natural selection processes, finding optimal solutions through iterations [84].

Predictive Simulations: Machine learning models can enhance predictive capabilities by approximating complex plasma dynamics, reducing computational costs, and enabling faster simulations.

- **Surrogate Models**: ML-based surrogate models approximate the behavior of detailed simulations, allowing for quick evaluations of plasma responses under various conditions.

- **Uncertainty Quantification**: Machine learning techniques can assess and reduce uncertainties in plasma simulations, providing more reliable predictions for complex systems [85].

Example of Machine Learning Application (Python):

```
import tensorflow as tf
from tensorflow import keras

# Load and preprocess plasma simulation data
data = ... # Load data from simulations
X_train, y_train = ... # Prepare training data

# Define a neural network model
model = keras.Sequential([
  keras.layers.Dense(128, activation='relu', input_shape=(X_train.shape[1],)),
  keras.layers.Dense(64, activation='relu'),
  keras.layers.Dense(1)
])

# Compile and train the model
model.compile(optimizer='adam', loss='mse')
model.fit(X_train, y_train, epochs=50, batch_size=32)

# Predict plasma behavior
predictions = model.predict(X_test)
```

6.2.5. Quantum Plasmas and Quantum Hydrodynamics

Quantum plasmas represent a state where quantum mechanical effects significantly influence plasma behavior. These effects become important at high densities and low temperatures, such as in astrophysical environments and advanced materials.

Quantum effects become significant in plasmas at extremely high densities and low temperatures, such as those found in white dwarfs, neutron stars, and certain laboratory conditions. Quantum plasma modelling incorporates quantum mechanical principles into traditional plasma models, providing a more accurate description of these extreme conditions. In quantum plasmas, effects such as quantum tunneling, wave-particle duality, and quantized energy levels play crucial roles.

- **Quantum Tunneling**: Particles can pass through potential barriers due to their wave-like nature, affecting collision processes and transport properties.

- **Quantized Energy Levels**: Discrete energy levels influence the absorption and emission spectra of plasmas, impacting radiation transport and energy balance [86].

In extreme conditions such as **astrophysical environments** or **high-density plasma systems**, **quantum effects** become significant. Traditional plasma models, which assume classical behaviors, fail to capture these effects accurately. **Quantum hydrodynamics** (QHD) extends classical fluid models by incorporating quantum mechanical principles, offering a more accurate framework for studying phenomena like **dense astrophysical plasmas**, **quantum materials**, and **novel states of matter** like **graphene** and **topological insulators**. Quantum plasma modeling is still in its early stages, but as quantum computing progresses, it will become increasingly important for understanding high-energy and high-density systems.

QHD extends classical hydrodynamics to include quantum effects, such as the Bohm potential, which accounts for quantum pressure. Quantum hydrodynamic models extend classical fluid descriptions to include quantum mechanical effects, providing a framework for studying quantum plasmas.

- **Governing Equations**: QHD equations incorporate terms for quantum pressure and Bohm potential, modifying the classical Navier-Stokes and MHD equations to account for quantum effects.
- **Applications**: Quantum hydrodynamics is used to model phenomena in dense astrophysical plasmas, semiconductor devices, and ultracold plasmas [87].

Quantum Kinetics: Uses quantum mechanical distribution functions and Wigner functions to describe particle dynamics in quantum plasmas. Quantum kinetic models describe the statistical behavior of particles in quantum plasmas, using tools like the Wigner function and quantum Boltzmann equation.

- **Wigner Function**: The Wigner function provides a phase-space representation of the quantum state, bridging the gap between classical and quantum descriptions.

- **Quantum Boltzmann Equation**: This equation models the time evolution of the distribution function, incorporating quantum effects in particle collisions and interactions [88].

Example of Quantum Plasma Simulation (C):

```
#include "quantum.h"
DEFINE_SOURCE(quantum_source, c, t, dS, eqn)
{
  real source = 0.0;
  real density = C_R(c, t);
  real quantum_pressure = ...; // Calculate quantum pressure

  /* Example source term for quantum hydrodynamics */
  source = quantum_pressure * density;

  dS[eqn] = source / density;

  return source;
}
```

6.3 The Future of Plasma Modeling

The future of plasma modeling lies in embracing and integrating emerging technologies and advanced topics that will revolutionize the field and enable the solving of more complex and larger-scale problems. As plasma science expands its influence across various disciplines and industries, innovations in modeling techniques and computational power will be crucial to unlocking new capabilities and applications. Here's a look at the key innovations driving plasma modeling forward, making it more efficient, comprehensive, and impactful.

This chapter explore how the convergence of emerging technologies, interdisciplinary research, and advanced computational methods are shaping the future of plasma modeling. However, this part offers more **practical insights** into the technologies that are currently advancing the field, with a focus on optimization, innovation, and real-world applications.

6.3.1. Integrated Multiphysics Modeling

One of the most significant advances in plasma modeling is the integration of **multiphysics models**. These models combine multiple physical processes, such as fluid dynamics, kinetic theory, quantum effects, and electromagnetism, to simulate the behavior of plasmas more accurately. When coupled with other fields like **solid mechanics** (to study plasma-material interactions) and **chemistry** (to simulate plasma-assisted chemical reactions), multiphysics modeling offers a holistic view of complex systems. This integration enables more precise predictions and optimizations, particularly in fields like **fusion energy**, **plasma-based manufacturing**, and **plasma medicine**.

Future innovations will focus on automating the coupling of these models, making simulations more user-friendly and adaptable to complex industrial problems. These models allow for more comprehensive simulations, capturing the interactions between plasma and other physical domains. For instance, integrating **plasma-material interactions** with **chemical reactions** can enhance the development of advanced coatings and surface treatments. This trend also involves cross-disciplinary collaboration, combining plasma science with **materials science**, **chemistry**, and **biomedical engineering** to solve complex, real-world problems.

6.3.2. Real-Time Simulation and Control

A game-changing development in plasma modeling is the use of **real-time simulation and control** systems. By integrating real-time data inputs from experiments with AI-driven simulations, these systems enable rapid adjustments and dynamic control of plasma processes. For example, in **fusion reactors**, real-time simulation can adjust magnetic confinement or plasma density to optimize performance and prevent instabilities. In **plasma medicine**, real-time control systems could tailor plasma treatments to individual patients, optimizing effectiveness while minimizing side effects. Such advancements will make plasma-based technologies more reliable, efficient, and adaptable to real-world conditions.

6.3.3. Plasma-Material Interactions

One of the most promising areas of innovation in plasma modeling is the integration of plasma science with **materials science**. Understanding how plasmas interact with materials at the atomic and molecular levels is crucial for optimizing processes like **plasma coating**, **surface modification**, and **etching**. By modeling these interactions in detail, engineers can develop more effective **plasma deposition** techniques, leading to advancements in industries such as **semiconductors**, **biomedical devices**, and **aerospace**. For example, plasma-assisted surface treatments can be optimized to create harder, more corrosion-resistant surfaces, while minimizing energy consumption and environmental impact.

6.3.4. Sustainable and Green Technologies

Plasma modeling will play a pivotal role in developing **sustainable technologies** aimed at addressing global environmental challenges. One promising area is the use of plasma for **waste treatment**, where plasma gasification can convert municipal waste into syngas, reducing landfill use and producing renewable energy. **Plasma-enhanced catalysis** and **plasma chemical processing** are also gaining attention as potential methods for reducing greenhouse gas emissions and creating **carbon-neutral fuels**. By advancing the efficiency of these processes through simulation, plasma modeling will contribute to the development of more environmentally friendly technologies.

6.3.5. Summary

The integration of **multiphysics modeling**, **real-time simulation**, **plasma-material interactions**, and **sustainable technologies** is driving the future of plasma science and its applications. **Integrated multiphysics models**, which combine fluid dynamics, kinetic theory, quantum effects, and electromagnetism, are enhancing our ability to simulate complex plasma behaviors, particularly in critical fields like **fusion energy**, **plasma-based manufacturing**, and **plasma medicine**. By coupling these models with other domains such as **solid mechanics** and **chemistry**, we can achieve more precise predictions and optimizations, leading to innovations in advanced coatings, surface treatments, and real-world industrial problems.

The use of **real-time simulation and control systems** promises to revolutionize plasma technology by enabling dynamic adjustments in processes like **fusion reactors** and **plasma**

medicine, improving reliability, efficiency, and adaptability. Furthermore, the increasing understanding of **plasma-material interactions** at the atomic and molecular levels is leading to more effective plasma deposition techniques and surface modifications, which are critical in industries like **semiconductors**, **biomedical devices**, and **aerospace**.

Additionally, plasma modeling is set to play a crucial role in the development of **sustainable and green technologies**. The application of **plasma gasification** for waste treatment, along with plasma-enhanced catalysis for reducing emissions and producing **carbon-neutral fuels**, offers significant environmental benefits. As plasma simulations continue to evolve, they will contribute to solving global challenges by advancing **renewable energy** and **environmentally friendly technologies**, marking a key step toward a more sustainable future.

6.4. Real-World Applications and Future Trends in Plasma Modelling

Plasma modelling has advanced significantly, driving innovation across various industries and research fields. This chapter highlights the real-world applications of plasma modelling, showcasing success stories and practical insights. It also explores future trends and innovations in computational methods, interdisciplinary integration, and addresses current challenges and opportunities in the field.

Real-World Applications of Plasma Modelling. Plasma modelling is critical in numerous industries, including semiconductor manufacturing, aerospace, energy, medicine, and environmental technology. Here, we highlight several success stories that demonstrate the practical impact of plasma modelling.

Source: Author

Figure 11. The illustration showcasing plasma technology across multiple industries: semiconductor manufacturing, aerospace, energy, medicine, and environmental technology.

6.4.1. Semiconductor Manufacturing:

Plasma etching and deposition processes are essential for creating microelectronic devices. Plasma modelling helps optimize these processes, ensuring high precision and uniformity.

- Success Story: Semiconductor companies have utilized plasma modelling to refine etching processes, achieving nanoscale accuracy and significantly reducing defects in integrated circuits. This optimization has led to faster, more reliable microchips, driving the advancement of modern electronics [89]

6.4.2. Aerospace Engineering:

Plasmas are used in propulsion systems and atmospheric re-entry processes. Modelling these plasmas helps improve performance and safety.

- **Success Story**: NASA and aerospace companies have employed plasma modelling to enhance the design of ion thrusters, which are used for long-duration space missions. These models have led to thrusters with higher efficiency and longer operational lifetimes, enabling deep-space exploration [90]

6.4.3. Energy Production:

Fusion energy research relies heavily on plasma modelling to understand and control high-temperature plasmas in reactors.

- **Success Story**: The ITER project, an international collaboration to develop a fusion reactor, uses plasma modelling to predict and manage plasma behavior. These models have been crucial in designing magnetic confinement systems that stabilize the plasma, bringing us closer to achieving sustainable fusion energy [91]

6.4.4. Medical Applications:

Non-thermal plasmas are used for sterilization, wound healing, and cancer treatment. Modelling helps optimize these applications for safety and efficacy.

- **Success Story**: Plasma medicine research has leveraged modelling to design devices that effectively sterilize surgical instruments and promote wound healing. These models have ensured that treatments are both safe and effective, leading to broader adoption in healthcare [92].

6.4.5. Environmental Technology:

Plasma processes are used for pollution control and waste treatment. Modelling these processes enhances their efficiency and effectiveness.

- **Success Story**: Plasma reactors have been developed to treat hazardous medical waste, using plasma modelling to optimize conditions for maximum disinfection and minimal emissions. These reactors are now used in hospitals and clinics, reducing the environmental impact of medical waste [93].

6.5. Current Challenges and Opportunities

Despite significant advancements, **plasma modeling** continues to face several challenges, each offering unique opportunities for future research and development.

Complexity and Nonlinearity: Plasmas exhibit highly nonlinear behavior across multiple scales, making accurate modeling difficult. The opportunity lies in developing **more advanced numerical methods and algorithms** to handle this complexity, which will improve our ability to predict and control plasma behavior in applications such as **fusion energy** and **plasma-based manufacturing**.

Validation and Verification: Ensuring that plasma models accurately reflect physical reality requires extensive validation against experimental data. The opportunity here is to establish **comprehensive validation frameworks** and **standardized experimental protocols**, improving the reliability of plasma models and their acceptance in industry and research.

Interdisciplinary Collaboration: Plasma modeling is inherently interdisciplinary, requiring collaboration across **physics**, **engineering**, **materials science**, and **computer science**. By fostering more effective interdisciplinary collaboration, there is an opportunity to develop innovative solutions and accelerate the progress of plasma technologies.

Scalability and Efficiency: Large-scale plasma simulations are computationally intensive, often requiring significant resources. Advancements in **high-performance computing (HPC)**, **cloud computing**, and **parallel processing** present opportunities to enhance the **scalability** and **efficiency** of these simulations, making them more accessible for a broader range of applications.

In conclusion, **plasma modeling** is crucial for driving innovation across industries like **semiconductor manufacturing**, **aerospace**, **fusion energy**, and **environmental technology**. Future trends, including **machine learning**, **quantum plasmas**, and **HPC**, will continue to address these challenges, expanding the potential and impact of plasma science and technology.

7: Thermal Plasma Modeling: A Theoretical Overview

Thermal plasma processing is a field that has seen significant growth and application in material science, metallurgy, and waste treatment. In the nineties, the rise of thermal plasma applications was highlighted due to its potential to achieve higher temperatures than conventional techniques, providing opportunities in melting, deposition, and material synthesis[94]. Thermal plasma modeling has become essential in understanding and optimizing these processes, as it allows for the simulation of various plasma characteristics, such as temperature distribution, particle behavior, and reaction kinetics. This article explores the key principles behind thermal plasma modeling and its relevance to plasma applications.

7.1 Key Principles of Thermal Plasma

Thermal plasma refers to a plasma in which all particles are in a near-equilibrium state, meaning the electrons and heavy particles (ions and neutrals) have similar temperatures. This equilibrium condition is known as Local Thermodynamic Equilibrium (LTE), where the electron temperature (T_e) is approximately equal to the heavy particle temperature (T_h). Thermal plasma is achieved at atmospheric or higher pressures, where collisions between particles are frequent enough to ensure energy exchange between electrons and ions.

Thermal plasmas are often generated using plasma torches, which create arcs between two electrodes. These torches are designed to produce a plasma jet that can reach temperatures in the range of 10,000 to 20,000 K. The high energy density of plasma allows for the processing of a variety of materials, including metals, ceramics, and hazardous waste. In industrial applications, thermal plasma is widely used for processes such as plasma arc welding, plasma spraying, and plasma gasification.

7.1.1. Plasma Modeling Fundamentals

Modeling thermal plasma involves simulating the behavior of plasma in response to external conditions, such as power input, gas flow, and material interaction. The primary goals of plasma modeling are to predict temperature distribution, flow dynamics, and species concentration, which are critical to optimizing plasma processes.

a. Governing Equations

- **Navier-Stokes Equations**: Plasma flow is governed by the Navier-Stokes equations, which describe the conservation of mass, momentum, and energy in a fluid. These equations must be adapted to account for the specific characteristics of plasma, such as ionization, recombination, and radiative heat transfer.
- **Energy Conservation**: The energy equation accounts for the heat generated by the plasma, which is primarily due to ohmic heating (the conversion of electrical energy into thermal energy) and radiative losses. In high-temperature plasmas, radiation can become a significant factor in energy transport.
- **Electromagnetic Fields**: Plasma is strongly influenced by electromagnetic fields. Maxwell's equations are used to describe the behavior of electric and magnetic fields in the plasma, including the generation of Lorentz forces that affect particle motion.

b. Temperature Distribution

One of the key challenges in plasma modeling is predicting the temperature distribution within the plasma jet. Thermal plasma is characterized by steep temperature gradients, particularly near the edges of the plasma jet. In most cases, the temperature in the plasma core can exceed 10,000 K, while the temperature near the boundary may drop significantly. Accurate modeling of temperature gradients is essential for predicting material behavior, especially in applications like plasma spraying, where the temperature distribution affects particle melting and deposition.

c. Plasma-Particle Interaction

In many thermal plasma applications, such as plasma spraying or waste treatment, particles are introduced into the plasma. These particles interact with the plasma through processes such as melting, evaporation, and chemical reactions. Plasma modeling must account for the heat transfer between the plasma and the particles, as well as the forces acting on the particles due to drag and Lorentz forces. Particle trajectories can be predicted by solving the equations of motion for the particles, considering both thermal and mechanical effects.

- **Local Thermodynamic Equilibrium (LTE) and Deviations** While thermal plasma is generally assumed to be in LTE, deviations from this assumption can occur, especially in low-pressure plasmas or near the edges of the plasma jet. In regions where LTE breaks down, the electron temperature may differ significantly from the heavy particle temperature. Plasma modeling must account for these deviations, as they can affect ionization and recombination rates, and hence the overall plasma behavior.

7.1.2. Applications of Thermal Plasma Modeling

Thermal plasma modeling has several key applications, including material processing, waste treatment, and synthesis of advanced materials. Some examples are discussed below:

a. **Plasma Melting and Remelting** Plasma melting is used in industries to melt metals and alloys at high temperatures. Plasma remelting offers advantages over traditional melting techniques, such as greater control over the melting environment and the ability to process refractory materials. Plasma modeling is crucial for predicting the temperature distribution in the melt pool and ensuring uniform melting.

b. **Plasma Spraying** In plasma spraying, particles are injected into a plasma jet and deposited onto a substrate to form a coating. The quality of the coating depends on the temperature and velocity of the particles, which in turn depend on the plasma characteristics. Plasma modeling is used to optimize the spray parameters, such as torch power, gas flow rate, and particle injection velocity, to achieve the desired coating properties.

c. **Plasma Gasification** Plasma gasification is an emerging waste treatment technology that uses thermal plasma to convert waste into syngas (a mixture of hydrogen and carbon monoxide). Plasma modeling helps optimize the gasification process by predicting the behavior of the plasma arc, the temperature distribution within the gasifier, and the interaction between the plasma and the waste feedstock. By fine-tuning these parameters, the efficiency of syngas production can be improved.

d. **Plasma Synthesis of Materials** Thermal plasma can be used to synthesize ultrafine powders and advanced materials. For example, plasma synthesis of nanomaterials has attracted significant attention due to the high quenching rates achievable in plasma systems, which prevent particle agglomeration and promote the formation of nanoscale structures. Plasma modeling is essential for understanding the reaction kinetics and optimizing the process parameters to achieve the desired material properties.

Thermal plasma modeling plays a critical role in the development and optimization of plasma-based technologies. By simulating the behavior of plasma and its interaction with materials, modeling allows researchers and engineers to predict process outcomes and make informed decisions regarding process parameters. As the field of thermal plasma processing continues to evolve, advancements in modeling techniques will be key to unlocking new applications and improving the efficiency of existing processes.

Thermal plasma processing has been a game-changing technology since its rise in the nineties, offering possibilities in material science, waste treatment, and metallurgy. As more advanced computational tools become available, thermal plasma modeling will continue to provide insights that drive innovation in these fields.

7.2 Modeling and Simulation of Plasma Melting and Remelting

Plasma melting and remelting are advanced material processing techniques used primarily in metallurgy for the purification and alloying of metals. These processes rely on thermal plasma to achieve temperatures far beyond those possible with traditional melting methods, allowing for efficient processing of high-temperature materials like refractory metals. To optimize these processes, modeling and simulation play a crucial role in predicting temperature distributions, energy transfer, and material behavior during the melting and remelting operations. This article delves into the theoretical background and simulation techniques used for modeling plasma melting and remelting processes.

7.2.1. Key Concepts in Plasma Melting and Remelting

Plasma melting refers to the use of plasma torches or arcs to melt materials in a furnace or crucible. Plasma remelting, on the other hand, involves the use of plasma to melt materials that have already been cast to improve their purity, uniformity, and microstructure. The main advantage of plasma remelting over conventional methods is the ability to control the furnace atmosphere, ensuring minimal contamination and allowing for precise control over the alloy composition.

The simulation of plasma melting and remelting requires the consideration of several important physical phenomena, including:

1. **Heat Transfer**: Plasma jets provide intense heat, which must be transferred uniformly throughout the molten material. Heat transfer occurs through conduction, convection, and radiation, and its simulation requires solving complex energy balance equations.
2. **Fluid Dynamics**: The molten pool in plasma melting exhibits fluid flow driven by temperature gradients

(thermocapillary forces), electromagnetic forces, and buoyancy effects. Modeling fluid flow is essential for understanding the distribution of heat and the uniformity of the melt.

3. **Electromagnetic Effects**: Plasma melting and remelting involve electric arcs or plasma torches, which generate electromagnetic fields. These fields interact with the molten material and induce forces that affect the flow of the molten metal, making the simulation of these effects critical.
4. **Phase Change and Solidification**: In remelting processes, the material undergoes phase changes from solid to liquid and then back to solid as it cools. Simulating these phase changes, including the solidification process, is essential for predicting the final microstructure and properties of the material.

7.2.2. Governing Equations for Plasma Melting Simulation

The modeling of plasma melting involves solving a set of partial differential equations (PDEs) that describe the behavior of the plasma and the material being processed. These equations include:

a. Conservation of Mass (Continuity Equation):

$$\frac{\partial \rho}{\partial t} - \nabla \cdot (\rho \mathbf{v}) = 0$$

Where ρ is the material density, v is the velocity field of the molten metal. This equation ensures that mass is conserved during the melting process.

b. Conservation of Momentum (Navier-Stokes Equation):

$$\rho \left(\frac{\partial \mathbf{v}}{\partial t} + \mathbf{v} \cdot \nabla \mathbf{v} \right) = -\nabla p + \mu \nabla^2 \mathbf{v} + \mathbf{F}_{EM}$$

Where p is the pressure, μ is the dynamic viscosity of the molten material, $\mathbf{F}_{EM}$ represents the electromagnetic forces acting on the fluid due to plasma arcs or torches. This equation models the fluid flow in the molten pool, driven by both thermal and electromagnetic forces.

c. Energy Conservation (Heat Transfer Equation):

$$\rho c_p \left(\frac{\partial T}{\partial t} + \mathbf{v} \cdot \nabla T \right) = k \nabla^2 T + \dot{Q}_{plasma}$$

Where c_p is the specific heat capacity of the material, T is the temperature, k is the thermal conductivity, Q_{plasma} is the heat input from the plasma arc or torch. This equation describes how heat is distributed throughout the molten material and how it evolves over time.

d. Maxwell's Equations (Electromagnetic Field):

$$\nabla \times \mathbf{E} = -\frac{\partial \mathbf{B}}{\partial t}, \quad \nabla \times \mathbf{B} = \mu_0 \mathbf{J}$$

Where E is the electric field, **B** is the magnetic field, **J** is the current density, μ_0 is the magnetic permeability. These equations describe the behavior of electromagnetic fields in the plasma arc and how they interact with the molten material.

7.2.3. Simulation Procedure

The simulation of plasma melting and remelting is a multi-step process that involves defining the initial conditions, solving the governing equations, and analyzing the results. A typical simulation procedure is outlined below:

1. **Define Initial Conditions**: The simulation begins by specifying the initial temperature distribution in the material, the initial flow field (if any), and the properties of the plasma arc (such as power, temperature, and velocity). The geometry of the furnace or crucible is also defined.

2. **Mesh Generation**: The simulation domain is discretized into a computational mesh or grid. The mesh resolution should be fine enough to capture steep temperature gradients near the plasma arc and the boundary layers in the molten pool.

3. **Solve Governing Equations**: The governing equations (continuity, momentum, energy, and Maxwell's equations) are solved iteratively using numerical methods, such as

finite element or finite volume methods. The solution provides the temperature distribution, velocity field, and electromagnetic field within the simulation domain.

4. **Incorporate Material Properties**: The material properties, such as thermal conductivity, specific heat, and viscosity, can change with temperature. These temperature-dependent properties are incorporated into the simulation to ensure accuracy.

5. **Simulate Phase Change**: In remelting processes, the simulation must account for the solid-liquid phase change. This can be done using a phase-change model, such as the enthalpy method, which tracks the latent heat released or absorbed during melting and solidification.

6. **Post-Processing and Analysis**: Once the simulation is complete, the results are analyzed to determine key parameters, such as the temperature distribution in the molten pool, the velocity of the molten metal, and the solidification front. These results help in optimizing the process parameters, such as plasma power, arc positioning, and gas flow rate.

a. Key Simulation Results

The results of plasma melting and remelting simulations provide valuable insights into the behavior of the molten material and the efficiency of the process. Some typical results include:

1. **Temperature Distribution**: The simulation reveals how heat is distributed within the molten pool. High temperatures are typically concentrated near the plasma arc, while the rest of the pool gradually heats up due to conduction and convection.

2. **Velocity Field**: The flow of molten metal is driven by thermocapillary forces and electromagnetic forces. The simulation helps predict the flow patterns, which are essential for ensuring uniform melting and mixing of the material.

3. **Solidification Front**: In remelting processes, the

simulation can track the solidification front as the material cools. This is important for predicting the final microstructure of the material and ensuring that the desired properties are achieved.

4. **Electromagnetic Field Distribution**: The simulation also provides the distribution of electromagnetic fields within the molten pool. This information is useful for optimizing the position of the plasma arc and adjusting the power input to achieve the desired melting behavior.

b. Applications of Plasma Melting and Remelting Simulation

Plasma melting and remelting simulations have a wide range of applications in industries that process high-performance materials, such as aerospace, automotive, and nuclear sectors. Some specific applications include:

1. **Production of Superalloys**: Plasma remelting is used to produce superalloys with precise chemical compositions and controlled microstructures. Simulation helps optimize the remelting process to achieve uniform alloying and minimize defects.

2. **Recycling of Scrap Metals**: Plasma melting is widely used in the recycling of scrap metals, such as aluminum and steel. Simulation allows for efficient melting of scrap materials while minimizing energy consumption and ensuring consistent material quality.

3. **Purification of Refractory Metals**: Plasma remelting is used to purify refractory metals, such as titanium and tungsten. Simulation helps control the remelting process to achieve high purity and uniformity in the final product.

Modeling and simulation are essential tools for optimizing plasma melting and remelting processes. By providing insights into temperature distribution, fluid dynamics, and electromagnetic effects, simulations enable engineers to fine-tune process parameters, improve material quality, and reduce energy consumption. As computational techniques continue to evolve, the accuracy and efficiency of plasma melting simulations will improve, paving the way for more advanced applications in material processing.

c. Simulation plasma melting and remelting using Python

To simulate **plasma melting and remelting** using Python, we can build a simplified model that incorporates some of the essential physics such as **heat transfer**, **melting**, and **remelting processes**. This model will simulate how a material changes its phase due to the heat generated by a plasma torch.

In this example, we'll use the **heat equation** to model the heat transfer within a solid material and simulate the melting process based on temperature thresholds. The key steps will be:

1. **Heat Transfer**: Model heat diffusion within the material due to the plasma's heat source.
2. **Phase Change**: Simulate phase change (melting/remelting) when the material's temperature exceeds its melting point.

We will use the **finite difference method** to solve the heat equation. Here's a simplified Python example for simulating plasma melting and remelting:

Python Code for Plasma Melting and Remelting Simulation

```
import numpy as np
import matplotlib.pyplot as plt

# Material properties
rho = 7800  # Density of the material (kg/m^3)
cp = 500  # Specific heat capacity (J/kg·K)
k = 45  # Thermal conductivity (W/m·K)
melting_point = 1500  # Melting point of the material (K)
latent_heat = 270000  # Latent heat of fusion (J/kg)

# Simulation parameters
L = 0.1  # Length of the material (m)
Nx = 100  # Number of spatial grid points
dx = L / (Nx - 1)  # Spatial resolution
alpha = k / (rho * cp)  # Thermal diffusivity
dt = 0.01  # Time step (s)
total_time = 50  # Total simulation time (s)

# Initialize temperature field
```

```
T = np.zeros(Nx)
T[int(Nx / 4):int(3 * Nx / 4)] = 300  # Initial temperature in the core
(K)

# Plasma torch heat source
plasma_power = 1e6  # Power of the plasma torch (W)
plasma_radius = 0.02  # Radius of the plasma heat source (m)
plasma_position = int(Nx / 2)  # Position of the plasma source
plasma_start = plasma_position - int(plasma_radius / dx)
plasma_end = plasma_position + int(plasma_radius / dx)

# Time-stepping loop
def simulate_melting_remelting():
    T_new = np.copy(T)
    heat_absorbed = np.zeros(Nx)  # Track latent heat absorption for
phase change

    for step in range(int(total_time / dt)):
        for i in range(1, Nx - 1):
            # Heat diffusion using finite difference
            T_new[i] = T[i] + alpha * dt / dx**2 * (T[i + 1] - 2 * T[i] + T[i -
1])

            # Add plasma torch heat source
            if plasma_start <= i <= plasma_end:
                T_new[i] += dt * plasma_power / (rho * cp *
plasma_radius)

            # Melting process (latent heat absorption)
            if T_new[i] >= melting_point and heat_absorbed[i] <
latent_heat:
                # Limit the temperature during melting by absorbing latent
heat
                heat_absorbed[i] += rho * latent_heat * dx
                T_new[i] = melting_point

        T[:] = T_new

        # Visualization update every 500 steps
        if step % 500 == 0:
            plt.plot(np.linspace(0, L, Nx), T, label=f'Time: {step*dt:.2f}s')

    plt.xlabel('Position (m)')
    plt.ylabel('Temperature (K)')
    plt.legend()
```

```
    plt.title('Plasma Melting and Remelting Simulation')
    plt.show()

# Run the simulation
simulate_melting_remelting()
```

Explanation of the Code:

1. **Material Properties**:
    - rho, cp, and k represent the **density**, **specific heat**, and **thermal conductivity** of the material, respectively.
    - melting_point is the temperature at which the material melts.
    - latent_heat represents the amount of energy required to change the material's phase from solid to liquid.
2. **Simulation Parameters**:
    - L is the length of the material being simulated.
    - Nx is the number of spatial points, and dx is the spatial resolution.
    - alpha is the thermal diffusivity, calculated using the thermal conductivity, density, and specific heat capacity.
3. **Heat Source**:
    - The plasma torch is simulated as a heat source applied at a specific location and over a specific radius.
4. **Melting and Remelting**:
    - If the material temperature exceeds the melting point, the model absorbs latent heat (the energy needed for phase change) and holds the temperature at the melting point until all latent heat is absorbed.
5. **Visualization**:
    - A plot is updated periodically to show the temperature

distribution across the material as it heats, melts, and potentially remelts.

How to Run This Simulation

To run this code:

1. Ensure you have Python installed.
2. Install numpy and matplotlib libraries via pip:

```
pip install numpy matplotlib
```

3. Run the Python script, and the simulation will generate a graph showing the temperature distribution across the material over time.

Extending the Model:

- **Cooling and Remelting**: You can extend this model to include cooling and remelting by adding additional phases where the heat source is turned off.
- **3D Simulation**: The model can be extended to 3D simulations by incorporating the heat equation in three dimensions.
- **Dynamic Heat Source**: The plasma torch can move over time, simulating a real-world scenario where the torch is scanned over the material.

This simplified model provides a basic framework for simulating **plasma melting** and **remelting** processes. By adjusting parameters and expanding the model, you can create more complex and realistic simulations tailored to specific industrial applications.

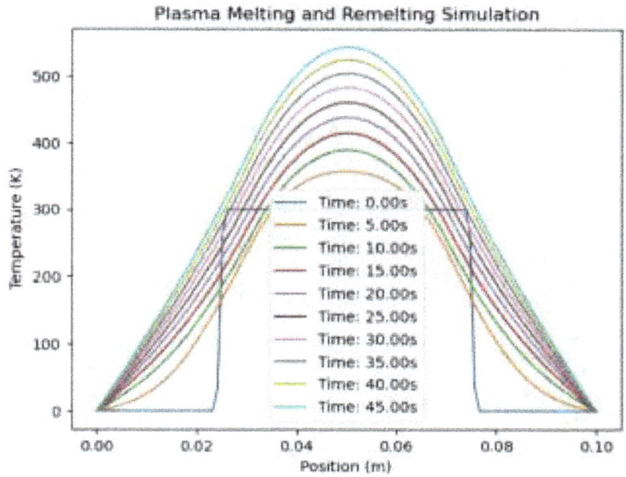

Figure 12. Temperature profile evolution plasma melting and remelting.

7.3 Modeling and Simulation of Plasma Spraying

Plasma spraying is a thermal spray process that uses plasma as the heat source to melt and propel particles onto a substrate, forming a coating. This process is widely employed in industries that require protective or functional coatings on components, such as aerospace, automotive, and biomedical sectors. The plasma spraying process can deposit materials like metals, ceramics, and polymers, offering high-performance coatings with desirable properties such as wear resistance, corrosion resistance, and thermal insulation.

To optimize plasma spraying, modeling and simulation are vital for predicting the behavior of the particles and plasma jet, understanding the coating formation, and enhancing the process efficiency. This article explores the theoretical framework behind plasma spraying and provides insights into the key factors modeled in the simulation.

7.3.1. Plasma Spraying Process

In the plasma spraying process, a plasma torch generates a high-velocity jet of plasma by ionizing a gas (usually argon, nitrogen, or hydrogen). The torch heats the gas to temperatures of up to 20,000 K, creating a high-energy plasma jet. Powder particles of the coating material are injected into the plasma jet, where they are heated, melted, and accelerated toward the substrate. Upon impact, the molten particles flatten and solidify, forming a layer of coating. Repeated impact of particles builds up the coating layer by layer.

Key variables in the plasma spraying process include:

- Plasma gas composition and flow rate
- Power input to the plasma torch
- Particle injection rate and velocity
- Stand-off distance (distance between torch and substrate)
- Substrate temperature

7.3.2. Key Concepts in Plasma Spraying Modeling

To simulate plasma spraying accurately, it is essential to model various physical processes that occur in the plasma jet and during particle deposition. These processes include fluid dynamics of the plasma jet, heat transfer, particle trajectory, melting, and solidification, as well as the coating formation on the substrate.

1. **Plasma Jet Dynamics**: The plasma jet is a high-velocity, high-temperature flow of ionized gas. The behavior of this jet, including temperature, velocity distribution, and interaction with the surrounding environment, is a key factor in determining particle heating and acceleration. The plasma jet is governed by fluid dynamics, and modeling this requires solving the Navier-Stokes equations for compressible flows.
2. **Particle Heating and Melting**: The particles injected into the plasma jet are rapidly heated by the plasma and, in most cases, fully melted before they reach the substrate. Heat transfer between the plasma and particles is influenced by particle size, velocity, and material properties. Simulation of particle heating requires solving

the energy conservation equation to track the temperature rise and phase change (from solid to liquid).

3. **Particle Trajectory and Acceleration**: Particles are subjected to various forces, including drag, buoyancy, and electromagnetic forces. These forces determine the trajectory and velocity of the particles as they move through the plasma jet. Accurate simulation of particle motion is necessary to ensure that the particles impact the substrate at the correct temperature and velocity for optimal coating formation.
4. **Coating Formation**: Upon impact with the substrate, the molten particles flatten and rapidly solidify, forming splats. These splats overlap to form the coating. The quality of the coating depends on the cooling rate, splat adhesion, and the porosity of the coating. Modeling the splat formation process requires understanding the dynamics of particle impact, cooling, and solidification.

7.3.3. Governing Equations for Plasma Spraying Simulation

The modeling of plasma spraying involves solving several key equations that describe the behavior of the plasma jet, particles, and coating formation:

1. **Navier-Stokes Equations (Fluid Dynamics of the Plasma Jet)**:

$$\frac{\partial \rho}{\partial t} + \nabla \cdot (\rho \mathbf{v}) = 0$$

$$\rho \left(\frac{\partial \mathbf{v}}{\partial t} + \mathbf{v} \cdot \nabla \mathbf{v} \right) = -\nabla p + \mu \nabla^2 \mathbf{v} + \mathbf{F}_{EM}$$

Where:

- ρ is the plasma density.
- v is the velocity of the plasma jet.
- μ is the dynamic viscosity.
- $\mathbf{F}_{EM}$ represents the electromagnetic forces.

- These equations govern the flow of the plasma jet, including the temperature and velocity profiles.

2. **Energy Conservation (Heat Transfer Equation for Particle Heating)**:

$$\rho_p c_p \left(\frac{\partial T_p}{\partial t} \right) = hA(T_{plasma} \quad T_p)$$

Where:

- ρ is the density of the particle.
- c_p is the specific heat capacity of the particle.
- *Tp* is the particle temperature.
- *Tplasma* is the temperature of the surrounding plasma.
- *h* is the heat transfer coefficient, and **A** is the surface area of the particle.
- This equation describes the heat transfer from the plasma to the particle, resulting in particle melting.

3. **Particle Trajectory (Newton's Second Law)**:

$$m_p \frac{d\mathbf{v}_p}{dt} = \mathbf{F}_{drag} + \mathbf{F}_{gravity} + \mathbf{F}_{electromagnetic}$$

Where:

- *mp* is the mass of the particle.
- *vp* is the velocity of the particle.
- **F***drag* represents the drag force exerted by the plasma on the particle.
- **F***gravity* is the gravitational force.
- **F***electromagnetic* accounts for electromagnetic forces (if present).
- This equation models the motion of the particles through the plasma jet.

7.3.4. Coating Formation (Splat Dynamics and Solidification):

The impact and solidification of molten particles on the substrate are modeled by tracking the flattening of the particle upon impact (splat formation) and the subsequent rapid cooling and solidification. This can be described using phase-change models (such as the Stefan problem) to capture the heat transfer during solidification:

$$\frac{\partial T}{\partial t} = \alpha \nabla^2 T$$

Where α is the thermal diffusivity.

7.3.5. Simulation Procedure for Plasma Spraying

The simulation of plasma spraying typically follows several steps:

1. **Define Plasma Torch and Particle Injection Parameters**: The first step is to define the characteristics of the plasma torch, such as power, gas composition, and flow rate. Particle properties, including size distribution, injection velocity, and mass flow rate, are also specified.

2. **Solve Plasma Jet Dynamics**: The plasma jet's velocity and temperature distribution are computed by solving the Navier-Stokes and energy conservation equations. This step provides the plasma conditions that affect particle heating and acceleration.

3. **Simulate Particle Motion and Heating**: The motion of individual particles is simulated by solving the equations of motion, considering drag, gravitational, and electromagnetic forces. Simultaneously, the heat transfer equation is solved to determine the particle's temperature and melting state.

4. **Coating Formation Simulation**: After particle impact, the splat formation and solidification are modeled to simulate the coating formation. This step helps determine the coating thickness, porosity, and microstructure.

5. **Post-Processing and Optimization**: The results of the simulation are analyzed to optimize process parameters such as torch power, gas flow, particle injection rate, and stand-off distance. This ensures that the desired coating properties (thickness, adhesion, porosity) are achieved.

a. Key Simulation Results

Simulation of plasma spraying provides valuable insights into various aspects of the process, including:

1. **Temperature and Velocity Profiles of the Plasma Jet**: The simulation reveals how temperature and velocity vary within the plasma jet. This helps in optimizing the plasma conditions to achieve the desired particle heating and acceleration.
2. **Particle Temperature and Melting State**: The simulation tracks the temperature evolution of particles as they move through the plasma jet, allowing engineers to adjust parameters for optimal particle melting before impact.
3. **Coating Thickness and Porosity**: The simulation predicts the final coating thickness and porosity, which are critical parameters for the performance of the coating. Higher particle velocities generally lead to denser coatings with lower porosity.
4. **Splat Formation and Microstructure**: The splat formation process is modeled to predict the coating's microstructure. Factors such as cooling rate and substrate temperature influence the adhesion strength and microstructural characteristics of the coating.

b. Applications of Plasma Spraying Simulation

Plasma spraying is widely used for applying protective and functional coatings in various industries. Key applications include:

1. **Thermal Barrier Coatings (TBCs)**: TBCs are used in gas turbines and aircraft engines to protect components from extreme temperatures. Plasma spraying is the preferred method for applying ceramic-based TBCs, and simulation helps optimize the coating's thermal insulation properties.

2. **Wear-Resistant Coatings**: Plasma spraying is used to apply coatings that improve wear resistance in industrial components, such as cutting tools, engine parts, and pumps. Simulation helps achieve the right balance between hardness and toughness.
3. **Biomedical Implants**: Plasma spraying is used to coat implants, such as hip joints and dental implants, with biocompatible materials like hydroxyapatite. Simulation ensures uniform coating thickness and excellent adhesion to the substrate.
4. **Corrosion-Resistant Coatings**: In industries like oil and gas, plasma spraying is used to apply corrosion-resistant coatings to protect equipment from harsh environments. Simulation helps optimize the coating's barrier properties.

Modeling and simulation of plasma spraying are essential for optimizing the process parameters and ensuring the production of high-quality coatings. By simulating the behavior of the plasma jet, particle heating, and coating formation, engineers can fine-tune the process to meet specific performance requirements. As computational tools continue to evolve, simulation will become an even more powerful tool for improving the efficiency and effectiveness of plasma spraying applications.

Source: Author

Figure 13. The illustration of vacuum tube plasma spraying applied to membrane coating, showcasing the precise and uniform application process.

c. Simulation of Plasma Spraying

Simulating **plasma spraying** involves modeling the interactions between a high-temperature plasma jet and the injected particles, capturing the behavior of the particles as they are melted and accelerated toward a substrate. Below is a simplified example of how we could implement a **plasma spraying simulation** using Python. This will model the temperature evolution and motion of particles in a plasma jet. For simplicity, we will assume a 1D model and ignore complex plasma jet dynamics like turbulence.

Assumptions for the model:

1. **Plasma jet**: Uniform temperature distribution.
2. **Particles**: Spherical, uniform material (e.g., metal).
3. **Heat transfer**: Conduction from plasma to particles.
4. **Motion**: Accelerated by drag from the plasma flow.

Python Implementation (simplified model)

```
mport numpy as np
import matplotlib.pyplot as plt

# Constants
plasma_temp = 15000  # K (plasma temperature)
initial_particle_temp = 300  # K (initial temperature of particles)
specific_heat = 500  # J/(kg*K) (specific heat capacity of particle
material)
density = 8000  # kg/m^3 (density of particle material)
particle_radius = 1e-4  # m (radius of the particles)
thermal_conductivity = 50  # W/(m*K) (thermal conductivity of
particle material)
plasma_velocity = 300  # m/s (velocity of plasma jet)
drag_coefficient = 0.5  # Drag coefficient
air_density = 1.225  # kg/m^3 (density of surrounding air)

# Simulation parameters
time_step = 0.0001  # s (time step for simulation)
total_time = 0.01  # s (total simulation time)
num_steps = int(total_time / time_step)
time = np.linspace(0, total_time, num_steps)

# Particle properties
mass = (4 / 3) * np.pi * (particle_radius ** 3) * density  # mass of the
particle
area = np.pi * particle_radius ** 2  # cross-sectional area of the
particle

# Arrays to store simulation results
particle_temp = np.zeros(num_steps)
particle_velocity = np.zeros(num_steps)
particle_position = np.zeros(num_steps)

# Initial conditions
particle_temp[0] = initial_particle_temp
```

```
particle_velocity[0] = 0
particle_position[0] = 0

# Heat transfer and drag calculation
def heat_transfer(temp_plasma, temp_particle):
    return thermal_conductivity * area * (temp_plasma -
temp_particle) / (density * specific_heat * particle_radius)

def drag_force(velocity_plasma, velocity_particle):
    relative_velocity = velocity_plasma - velocity_particle
    return 0.5 * air_density * drag_coefficient * area * relative_velocity
** 2

# Simulation loop
for i in range(1, num_steps):
    # Calculate heat transfer and update temperature
    heat_rate = heat_transfer(plasma_temp, particle_temp[i-1])
    particle_temp[i] = particle_temp[i-1] + heat_rate * time_step

    # Calculate drag force and update velocity
    drag = drag_force(plasma_velocity, particle_velocity[i-1])
    particle_velocity[i] = particle_velocity[i-1] + (drag / mass) *
time_step

    # Update position of the particle
    particle_position[i] = particle_position[i-1] + particle_velocity[i] *
time_step

# Plotting results
plt.figure(figsize=(10, 6))

# Plot temperature
plt.subplot(2, 1, 1)
plt.plot(time, particle_temp)
plt.title("Particle Temperature During Plasma Spraying")
plt.xlabel("Time (s)")
plt.ylabel("Temperature (K)")
plt.grid(True)

# Plot particle velocity
plt.subplot(2, 1, 2)
plt.plot(time, particle_velocity)
plt.title("Particle Velocity During Plasma Spraying")
plt.xlabel("Time (s)")
plt.ylabel("Velocity (m/s)")
```

```
plt.grid(True)

plt.tight_layout()
plt.show()
```

Key Features of the Model:

1. **Heat Transfer**: Modeled using Fourier's law, this code computes how quickly a particle heats up in a plasma jet, based on the temperature difference between the particle and the plasma.
2. **Particle Acceleration**: The code computes the velocity of the particle as it is accelerated by drag forces from the plasma.
3. **Position Tracking**: As the particle is accelerated, the model tracks its movement over time.

Model Outputs:

- **Particle Temperature Over Time**: This shows how the particle heats up as it travels through the plasma jet.
- **Particle Velocity Over Time**: This indicates how quickly the particle is accelerated by the drag force from the plasma jet.

Enhancements:

This is a simplified model. More realistic simulations would require:

1. **2D or 3D modeling**: To simulate the full spatial dynamics of the plasma and particle interactions.
2. **Turbulence modeling**: Incorporating plasma turbulence and instabilities.
3. **Variable particle size distribution**: To account for real-world scenarios where particles may have different sizes and properties.

This basic framework, however, provides a starting point for more sophisticated plasma spraying simulations.

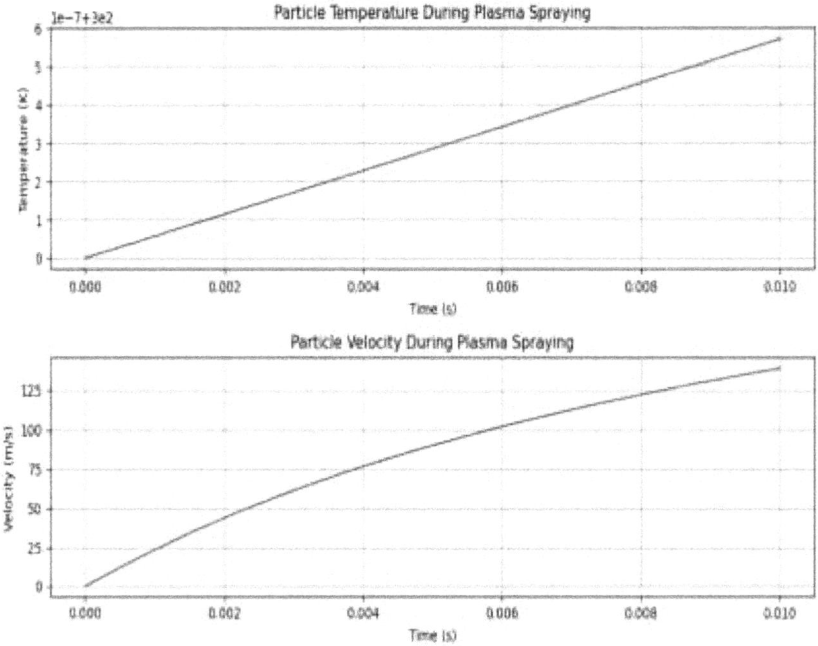

Figure 14. Model output of plasma spraying. Particle Temperature Over Time (above) and Particle Velocity Over Time (below)

7.4 Modeling and Simulation of Plasma Synthesis of Materials

Introduction

Plasma synthesis of materials is an advanced method for producing nanomaterials, ultrafine powders, and complex compounds that are difficult or impossible to synthesize using traditional chemical or thermal methods. The plasma environment provides the unique ability to achieve extremely high temperatures and rapid cooling rates, which are critical for controlling the properties of the synthesized materials, such as particle size, phase composition, and morphology. Plasma synthesis has become a cornerstone in various industries, including electronics, pharmaceuticals, and materials science, for producing high-purity, high-performance materials.

Modeling and simulation play a crucial role in optimizing plasma synthesis processes by predicting the interactions between plasma, precursor materials, and the resulting particles.

In this article, we explore the theoretical framework and computational techniques used to model plasma synthesis and discuss its key applications.

7.4.1. Plasma Synthesis Process

Plasma synthesis relies on the generation of a high-temperature plasma, typically through methods such as radio-frequency (RF) plasma, direct-current (DC) plasma, or microwave plasma. In the plasma synthesis process, precursor materials, which may be in the form of gas, liquid, or solid, are introduced into the plasma region. The intense energy of the plasma causes the precursor to dissociate and ionize, leading to the formation of reactive species (atoms, ions, and radicals). These species then undergo nucleation, growth, and agglomeration to form the final material, which is collected as ultrafine powder or nanomaterial.

Key steps in plasma synthesis include:

- **Dissociation**: The precursor material is dissociated into reactive species by the plasma energy.
- **Nucleation and Growth**: The reactive species undergo nucleation, forming seed particles, which then grow by aggregation or condensation of additional atoms or molecules.
- **Agglomeration and Quenching**: The plasma environment allows rapid particle growth, and the synthesized particles are rapidly cooled to prevent further agglomeration or phase transformation.

The plasma synthesis process can be finely tuned by controlling factors such as plasma temperature, precursor feed rate, gas flow rate, and residence time in the plasma zone. Simulation is used to optimize these parameters and ensure the desired material properties are achieved.

7.4.2. Key Concepts in Plasma Synthesis Modeling

Modeling the plasma synthesis process involves simulating various physical phenomena, such as plasma chemistry, heat transfer, fluid dynamics, and particle dynamics. Key factors to consider in the modeling of plasma synthesis include:

1. **Plasma Chemistry**: The plasma environment provides a unique chemical reactor where ionization, dissociation, and recombination reactions occur. Modeling plasma chemistry involves solving reaction kinetics equations to predict the formation of reactive species and their interaction with precursor materials.
2. **Nucleation and Particle Growth**: Particle nucleation occurs when atoms or molecules cluster together to form the initial seed particles. This process is followed by particle growth, which can occur through condensation, coagulation, or surface growth. The modeling of nucleation and growth requires tracking the size distribution of particles and their interaction with the plasma species.
3. **Thermal and Fluid Dynamics**: The plasma flow plays a significant role in transporting precursor materials and growing particles. Simulating the fluid dynamics of the plasma helps in understanding how particles are carried through the plasma zone, how they interact with the gas flow, and how heat is transferred between the plasma and the particles.
4. **Quenching and Cooling**: Rapid cooling or quenching is essential to control particle size and prevent unwanted phase transformations. The modeling of quenching involves solving the energy balance equations to predict the temperature drop of particles as they exit the plasma region.
5. **Phase Transformation**: Plasma synthesis can produce materials with complex phase compositions, such as metastable phases or amorphous structures. The simulation of phase transformations requires understanding the thermodynamics and kinetics of the material system to predict how different phases form and evolve under plasma conditions.

7.4.3. Governing Equations for Plasma Synthesis Simulation

The modeling of plasma synthesis involves solving a set of partial differential equations (PDEs) that describe the behavior of the plasma, the precursor materials, and the synthesized particles. Key equations include:

1. **Plasma Flow Dynamics (Navier-Stokes Equations)**:

$$\frac{\partial \rho}{\partial t} - \nabla \cdot (\rho \mathbf{v}) = 0$$

$$\rho\left(\frac{\partial \mathbf{v}}{\partial t} + \mathbf{v} \cdot \nabla \mathbf{v}\right) = -\nabla p + \mu \nabla^2 \mathbf{v} + \mathbf{F}_{EM}$$

Where:

- ρ is the plasma density.
- v is the velocity of the plasma flow.
- p is the pressure.
- μ is the dynamic viscosity.
- $\mathbf{F}_{\mathrm{EM}}$ represents electromagnetic forces acting on the plasma.

2. **Energy Conservation (Heat Transfer Equation)**:

$$\rho c_p\left(\frac{\partial T}{\partial t} + \mathbf{v} \cdot \nabla T\right) = k \nabla^2 T + Q_{plasma}$$

Where:

- $\boldsymbol{T}$ is the temperature of the plasma or particles.
- c_p is the specific heat capacity.
- K is the thermal conductivity.
- Q_{plasma} represents heat input from the plasma to the particles.

3. **Particle Growth Dynamics (Population Balance Equation)**:

$$\frac{\partial f(v,t)}{\partial t} + \frac{\partial}{\partial v}\left(G(v) f(v,t)\right) = R_n(v,t)$$

Where:

- $f_{(v,t)}$ is the particle size distribution function.
- v is the particle volume.
- $G_{(v)}$ represents the particle growth rate.
- R_n(v,t) is the nucleation rate of particles.

4. **Reaction Kinetics for Plasma Chemistry**:

$$\frac{dC_i}{dt} \quad \sum_j k_{ij} C_j \quad \sum_k k_{ik} C_i$$

Where:

- C_i is the concentration of species iii.
- k_{ij} is the rate constant for the reaction between species *i* and *j*.
- This equation tracks the formation and depletion of reactive species in the plasma.

7.4.4. Simulation Procedure for Plasma Synthesis

A typical simulation of plasma synthesis involves the following steps:

1. **Define Plasma and Precursor Parameters**: The first step is to define the characteristics of the plasma, such as gas composition, power input, and flow rate. Precursor materials are also defined in terms of their composition, feed rate, and initial particle size.
2. **Solve Plasma Dynamics and Chemistry**: The plasma flow and chemical reactions are simulated using fluid dynamics and reaction kinetics models. This step predicts the temperature and concentration distribution of reactive species within the plasma.
3. **Model Particle Nucleation and Growth**: The nucleation and growth of particles are simulated by solving the population balance equation. The size distribution of particles and their evolution over time are tracked throughout the process.
4. **Simulate Particle Quenching and Collection**: As

particles exit the plasma zone, their temperature rapidly decreases. This quenching process is simulated to ensure that the desired particle size and phase are achieved.

5. **Post-Processing and Optimization**: The simulation results are analyzed to optimize the plasma synthesis parameters, such as residence time, precursor feed rate, and plasma power, to achieve the desired material properties.

a. Key Simulation Results

Simulation of plasma synthesis provides valuable insights into the behavior of the plasma and the resulting materials. Some typical results include:

1. **Temperature and Concentration Distribution**: The simulation reveals the temperature and concentration profiles within the plasma, helping optimize the conditions for precursor dissociation and particle formation.
2. **Particle Size Distribution**: The simulation tracks the evolution of particle size, allowing engineers to adjust process parameters to achieve the desired particle size distribution for the final material.
3. **Phase Composition**: The simulation predicts the phase composition of the synthesized material, including the formation of metastable phases or amorphous structures.
4. **Nucleation and Growth Rates**: The rates of nucleation and growth of particles are key indicators of the efficiency of the plasma synthesis process. Simulation provides insights into how these rates can be controlled by adjusting plasma conditions.

b. Applications of Plasma Synthesis Simulation

Plasma synthesis is used to produce a wide range of materials with tailored properties. Key applications include:

1. **Nanoparticle Synthesis**: Plasma synthesis is widely used to produce nanoparticles with controlled size, morphology, and phase composition. Simulation helps

optimize the process for applications in electronics, catalysis, and medicine.

2. **Advanced Ceramics**: Plasma synthesis is used to produce advanced ceramics with high purity and tailored microstructures. Simulation ensures that the desired phase composition and mechanical properties are achieved.
3. **Powder Metallurgy**: Plasma synthesis is employed to produce ultrafine metal powders for applications in powder metallurgy and additive manufacturing. Simulation helps control particle size and minimize agglomeration.
4. **Composite Materials**: Plasma synthesis can produce composite materials by co-synthesizing different components. Simulation is used to optimize the distribution and interaction of different phases within the composite.

Modeling and simulation are essential tools for optimizing plasma synthesis processes. By simulating the behavior of plasma, precursor materials, and growing particles, engineers can fine-tune the process parameters to achieve high-quality materials with the desired properties. As plasma synthesis continues to evolve, advancements in simulation techniques will play a critical role in enabling the production of next-generation materials for a wide range of industries.

c. Simplified simulation materials synthesis

To simulate **Plasma Synthesis of Materials**, we can create a simplified Python model that focuses on key aspects such as **particle heating**, **nucleation**, and **growth** in the plasma environment. The goal is to simulate the evolution of particle temperature and size as they interact with the plasma and grow through condensation or aggregation. We will assume basic principles of heat transfer and particle kinetics. Below is a Python-based simulation using simplified physical assumptions:

Python Simulation: Plasma Synthesis of Nanoparticles

This simulation tracks the temperature and size of nanoparticles during plasma synthesis using basic heat transfer equations and particle growth dynamics. For simplicity, we use a heat transfer equation to model particle heating and a basic growth rate equation for nucleation and growth.

```
import numpy as np
import matplotlib.pyplot as plt

# Constants
T_plasma = 10000  # Plasma temperature in K
T_initial = 300   # Initial temperature of particles in K
Cp = 500          # Specific heat capacity of particles (J/kg K)
particle_mass = 1e-9  # Mass of individual nanoparticles (kg)
h = 100  # Heat transfer coefficient (W/m^2K)
area = 1e-12  # Surface area of the particle (m^2)
growth_rate = 1e-9  # Growth rate of particle size (m/s)
dt = 0.1  # Time step in seconds
time_total = 10  # Total simulation time in seconds

# Arrays to store results
time_steps = np.arange(0, time_total, dt)
temperature = np.zeros(len(time_steps))
size = np.zeros(len(time_steps))

# Initial conditions
temperature[0] = T_initial
size[0] = 1e-9  # Initial particle size (m)

# Simulation loop
for i in range(1, len(time_steps)):
   # Heat transfer to the particle
   dT = (h * area * (T_plasma - temperature[i-1])) / (particle_mass *
Cp) * dt
   temperature[i] = temperature[i-1] + dT

   # Particle growth rate (simplified linear growth)
   size[i] = size[i-1] + growth_rate * dt

# Plotting results
plt.figure(figsize=(12, 6))

plt.subplot(1, 2, 1)
plt.plot(time_steps, temperature, label="Particle Temperature (K)")
```

```
plt.xlabel("Time (s)")
plt.ylabel("Temperature (K)")
plt.title("Particle Temperature Evolution in Plasma")
plt.grid(True)

plt.subplot(1, 2, 2)
plt.plot(time_steps, size * 1e9, label="Particle Size (nm)")
plt.xlabel("Time (s)")
plt.ylabel("Size (nm)")
plt.title("Particle Size Growth in Plasma")
plt.grid(True)

plt.tight_layout()
plt.show()
```

Key Components of the Simulation:

1. **Particle Heating**:
   - The heat transfer to the particle is modeled using a simplified heat transfer equation:

     $$\frac{dT}{dt} = \frac{hA(T_{plasma} - T_{particle})}{mC_p}$$

     Here, h is the heat transfer coefficient, A is the surface area of the particle, and T_plasma is the plasma temperature.

2. **Particle Growth**:
   - The particle size increases over time as the particles nucleate and grow within the plasma. This growth is modeled using a simple linear growth rate for simplicity:

     $$\frac{dR}{dt} = G$$

   - Where G is the growth rate of the particle size.

Results:

- **Temperature Evolution**: The particle temperature increases over time as it absorbs heat from the plasma.
- **Particle Growth**: The particle size grows linearly over time due to condensation or aggregation processes within

the plasma.

This simplified model can be further extended to include more complex factors like multi-step nucleation, detailed growth kinetics, and plasma chemistry.

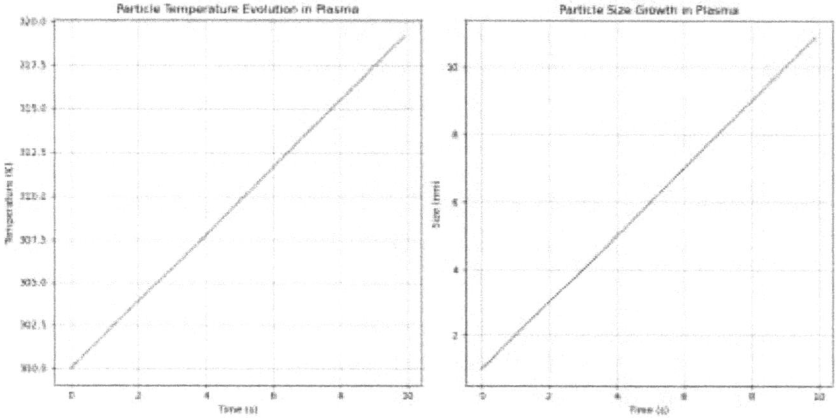

Figure 15. The heat transfer to the particle (left) and the particle size increases over time as the particles nucleate and grow within the plasma (right).

7.5 Modeling and Simulation of Plasma Gasification

Plasma gasification is an advanced waste treatment technology that converts carbon-containing materials, such as municipal solid waste (MSW), biomass, and hazardous waste, into syngas (a mixture of hydrogen and carbon monoxide) through the application of high-temperature plasma. This process is gaining traction due to its ability to handle a wide variety of feedstocks, including waste that is difficult to treat using conventional methods, and its potential for energy recovery. By transforming waste into useful energy sources, plasma gasification presents an environmentally friendly solution for both waste management and renewable energy production.

Modeling and simulation play a critical role in the development and optimization of plasma gasification technologies by predicting the behavior of the plasma, the feedstock, and the resulting syngas composition. In this article, we will explore the theoretical framework behind plasma gasification and discuss the key factors involved in its modeling and simulation.

7.5.1. Plasma Gasification Process

In plasma gasification, a high-temperature plasma torch (reaching temperatures between 3,000°C and 7,000°C) breaks down the feedstock into its elemental components. The plasma torch provides the necessary energy to ionize the feedstock, dissociating complex molecules into simpler gases and breaking down inorganic materials into vitrified slag.

Key stages of the plasma gasification process include:

- **Feedstock Preparation**: MSW or other feedstocks are pre-processed, shredded, and dried to ensure uniformity and reduce moisture content.
- **Plasma Gasification**: The feedstock is exposed to the plasma arc in an oxygen-starved environment, where the high temperatures cause thermal decomposition of organic materials into syngas.
- **Syngas Cleaning**: The syngas produced is cleaned to remove contaminants such as tars, particulates, and heavy metals, making it suitable for further use.
- **Energy Recovery**: The cleaned syngas can be used to generate electricity, produce heat, or be further processed into synthetic fuels.

The main products of plasma gasification are syngas and vitrified slag, both of which can be utilized, making plasma gasification an attractive option for waste-to-energy applications. To optimize the process, modeling and simulation are used to predict how the feedstock interacts with the plasma, how the syngas composition evolves, and how energy recovery can be maximized.

7.5.2. Key Concepts in Plasma Gasification Modeling

The modeling of plasma gasification involves simulating several physical and chemical processes, including plasma generation, heat and mass transfer, chemical reactions, and feedstock decomposition. Key factors to consider in plasma gasification modeling include:

1. **Plasma Torch Dynamics**: The plasma torch provides the energy required for gasification. Simulating the behavior of the plasma torch, including the temperature and velocity distribution of the plasma jet, is crucial for understanding how the energy is transferred to the feedstock. The plasma jet interacts with the feedstock and drives the decomposition of organic materials.
2. **Feedstock Decomposition**: The organic materials in the feedstock are thermally decomposed into simpler molecules, primarily hydrogen (H_2), carbon monoxide (CO), and smaller amounts of methane (CH_4). The rate of decomposition and the resulting gas composition depend on the feedstock properties, temperature, and residence time in the plasma zone. Modeling these factors is essential for predicting syngas composition.
3. **Chemical Reactions in the Plasma**: Plasma gasification involves complex chemical reactions, including ionization, dissociation, and recombination. These reactions take place in the plasma arc and gasifier. Modeling plasma chemistry is essential for understanding how different feedstock components react and how syngas composition evolves.
4. **Heat and Mass Transfer**: Heat transfer from the plasma to the feedstock is a key factor in determining the efficiency of gasification. Accurate modeling of heat and mass transfer processes is necessary to ensure complete decomposition of the feedstock and efficient production of syngas.
5. **Slag Formation**: Inorganic components of the feedstock (e.g., metals, glass, and minerals) are melted in the plasma and form a vitrified slag. This slag can be removed from the gasifier and used in construction materials or disposed of safely. Modeling slag formation and its interaction with the plasma is important for predicting the efficiency of

waste conversion and the quality of the slag.

7.5.3. Governing Equations for Plasma Gasification Simulation

The modeling of plasma gasification involves solving several key equations that describe the behavior of the plasma, the feedstock, and the syngas production. These equations include:

1. **Plasma Flow Dynamics (Navier-Stokes Equations)**
2. **Energy Conservation (Heat Transfer Equation)**
3. **Chemical Reaction Kinetics**:

   $$\frac{dC_i}{dt} = \sum_j k_{ij} C_j - \sum_k k_{ik} C_i$$

   Where:

   - C_i is the concentration of species *i*.
   - k_{ij} is the rate constant for the reaction between species iii and jjj.
   - This equation tracks the formation of syngas components (H_2, CO, CH_4, etc.) and other byproducts.

4. **Mass Transfer for Feedstock Decomposition**:

   $$\frac{\partial C_f}{\partial t} + \mathbf{v} \cdot \nabla C_f = -r_f$$

   Where:

   - C_f is the concentration of feedstock components.
   - r_f is the rate of feedstock decomposition.
   - This equation models the breakdown of the feedstock as it interacts with the plasma.

7.5.4. Simulation Procedure for Plasma Gasification

A typical simulation of plasma gasification involves the following steps:

1. **Define Plasma Torch and Feedstock Parameters**: The

first step is to define the characteristics of the plasma torch, including power, gas composition, and flow rate. The properties of the feedstock, such as moisture content, particle size, and composition, are also specified.

2. **Solve Plasma Flow and Temperature Distribution**: The plasma flow and temperature distribution are computed by solving the Navier-Stokes and energy conservation equations. This step provides the temperature and velocity profiles in the plasma jet and gasifier.
3. **Model Feedstock Decomposition and Syngas Production**: The decomposition of the feedstock is simulated using mass transfer and chemical reaction kinetics models. The composition of the syngas is tracked as the feedstock interacts with the plasma.
4. **Simulate Slag Formation and Removal**: The formation of slag is modeled based on the melting and fusion of inorganic materials in the feedstock. The removal and quality of the slag are predicted to optimize waste-to-energy conversion.
5. **Post-Processing and Optimization**: The results of the simulation are analyzed to optimize process parameters such as plasma power, feedstock feed rate, and gas flow rate to maximize syngas production and energy efficiency.

a. Key Simulation Results

Simulation of plasma gasification provides insights into various aspects of the process, including:

1. **Temperature and Velocity Profiles**: The simulation reveals the temperature and velocity distribution within the plasma jet and gasifier. These profiles help optimize the energy transfer to the feedstock and the overall efficiency of the gasification process.
2. **Syngas Composition**: The simulation predicts the composition of the syngas, including the concentration of hydrogen, carbon monoxide, and other gases. This information is crucial for optimizing the gas cleaning

process and energy recovery.

3. **Feedstock Decomposition Rates**: The simulation tracks the decomposition of the feedstock and the rate at which syngas is produced. This helps determine the optimal feedstock feed rate and residence time in the gasifier.
4. **Slag Formation and Quality**: The simulation predicts the quantity and quality of the slag produced. This information is useful for minimizing waste and ensuring that the slag can be safely used or disposed of.

b. Applications of Plasma Gasification Simulation

Plasma gasification is used for a variety of applications, including waste-to-energy conversion, hazardous waste treatment, and biomass gasification. Key applications include:

1. **Municipal Solid Waste (MSW) Treatment**: Plasma gasification is used to convert MSW into syngas, which can be used to generate electricity or produce synthetic fuels. Simulation helps optimize the process to maximize energy recovery and minimize emissions.
2. **Hazardous Waste Disposal**: Plasma gasification is effective in treating hazardous wastes, such as medical waste or electronic waste, by converting toxic materials into inert slag. Simulation helps ensure complete destruction of hazardous components.
3. **Biomass Gasification**: Biomass feedstocks, such as agricultural residues or wood chips, can be gasified using plasma to produce renewable syngas. Simulation helps optimize the process for different types of biomass and ensures high energy efficiency.
4. **Industrial Waste Treatment**: Plasma gasification is used to treat industrial wastes, such as slag and ash from metal production or chemical waste. Simulation helps optimize the conversion of these wastes into useful energy products.

Modeling and simulation are essential tools for optimizing plasma gasification processes. By simulating the behavior of the plasma, the feedstock, and the resulting syngas composition, engineers can fine-tune the process parameters to achieve maximum efficiency and environmental sustainability. As plasma gasification continues to evolve, advancements in simulation techniques will be key to expanding its applications in waste treatment and energy production.

c. Python simulation

To simulate **plasma gasification**, we will need to consider some essential components, including **energy balance**, **reaction kinetics**, and **syngas composition** from the gasification process. Below is a Python-based simulation for a simplified plasma gasification model that calculates syngas composition and energy distribution based on a set of predefined parameters.

Here's an example of a basic plasma gasification simulation in Python:

```
import numpy as np
import matplotlib.pyplot as plt

# Constants
gasification_temp = 3000  # Plasma temperature in Kelvin
R = 8.314  # Universal gas constant, J/(mol*K)
feedstock_mass = 100  # Mass of feedstock (kg)
carbon_fraction = 0.5  # Carbon fraction in feedstock
water_fraction = 0.3  # Water fraction in feedstock
inert_fraction = 0.2  # Inert fraction in feedstock (non-reactive)

# Reaction constants
carbon_gasification_rate = 1e-2  # Rate of carbon gasification reaction
(mol/s)
water_gas_shift_rate = 1e-3  # Rate of water-gas shift reaction (mol/s)

# Time parameters for simulation
time_steps = 1000
time = np.linspace(0, 1000, time_steps)  # Time from 0 to 1000
seconds

# Gas composition
CO = np.zeros(time_steps)  # Carbon monoxide concentration
H2 = np.zeros(time_steps)  # Hydrogen concentration
CO2 = np.zeros(time_steps)  # Carbon dioxide concentration
```

```
# Reaction equations
def gasification(carbon, water, temp):
   # Simplified gasification reaction: C + H2O -> CO + H2
   gas_yield = carbon_gasification_rate * carbon * np.exp(-1000 /
temp)
   CO_produced = gas_yield
   H2_produced = gas_yield
   return CO_produced, H2_produced

def water_gas_shift(CO_conc, water):
   # Water-gas shift reaction: CO + H2O -> CO2 + H2
   reaction_yield = water_gas_shift_rate * CO_conc
   CO2_produced = reaction_yield
   H2_additional = reaction_yield
   CO_reduced = reaction_yield
   return CO2_produced, H2_additional, CO_reduced

# Initialize concentrations
carbon_content = feedstock_mass * carbon_fraction  # Carbon mass in
feedstock
water_content = feedstock_mass * water_fraction  # Water mass in
feedstock

# Simulate gasification over time
for t in range(1, time_steps):
   CO_produced, H2_produced = gasification(carbon_content,
water_content, gasification_temp)
   CO2_prod, H2_add, CO_red = water_gas_shift(CO[t-1],
water_content)

   # Update gas concentrations
   CO[t] = CO[t-1] + CO_produced - CO_red
   H2[t] = H2[t-1] + H2_produced + H2_add
   CO2[t] = CO2[t-1] + CO2_prod

   # Reduce carbon content as it reacts
   carbon_content -= CO_produced
   water_content -= (CO_produced + CO2_prod)

# Plotting results
plt.figure(figsize=(10, 6))
plt.plot(time, CO, label='CO concentration (mol)')
plt.plot(time, H2, label='H2 concentration (mol)')
plt.plot(time, CO2, label='CO2 concentration (mol)')
```

```
plt.xlabel('Time (s)')
plt.ylabel('Concentration (mol)')
plt.title('Plasma Gasification Syngas Composition Over Time')
plt.legend()
plt.grid(True)
plt.show()
```

Explanation:

1. **Feedstock Composition**: We define the mass of the feedstock and its carbon, water, and inert fractions. This model assumes that the plasma will only gasify the carbon and water portions.
2. **Gasification Reaction**: The first reaction in the model is the basic gasification of carbon with water to form carbon monoxide (CO) and hydrogen (H_2).
3. **Water-Gas Shift Reaction**: After gasification, the produced CO undergoes the water-gas shift reaction to produce additional hydrogen and carbon dioxide (CO_2).
4. **Reaction Rates**: Reaction rates are simplified as exponential functions of the temperature, representing how the gasification reactions proceed with time.
5. **Results**: The code simulates how the concentrations of CO, H_2, and CO_2 evolve over time, providing insights into the syngas composition.

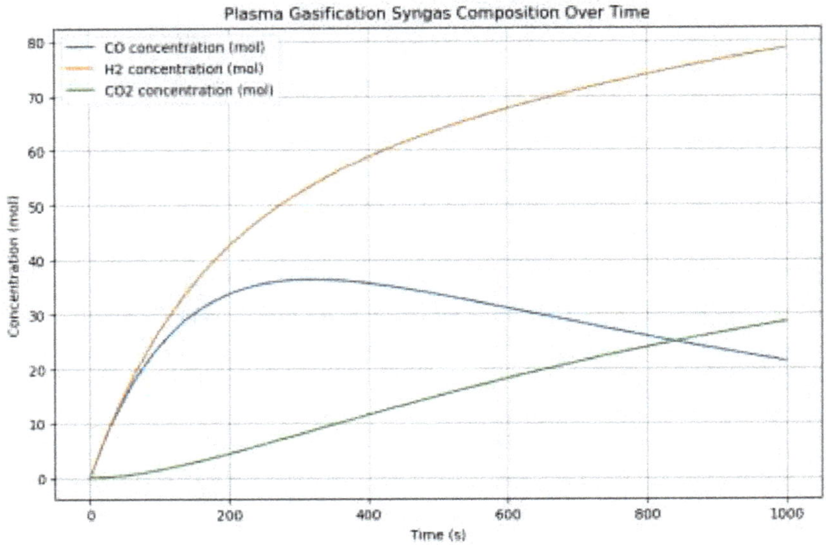

Figure 16. The plot shows how the concentrations of CO, H_2, and CO_2 evolve over time from the plasma gasification process.

Visualization:

The plot shows how the concentrations of CO, H_2, and CO_2 evolve over time, allowing you to visualize the syngas composition produced from the plasma gasification process.

This simulation can be further expanded by refining the reaction rates, adding more detailed thermodynamic considerations, and simulating heat transfer in the gasification reactor.

8: Plasma Gasifiers

Plasma gasifiers represent an innovative technology for converting waste materials into valuable synthesis gas (syngas) through the use of plasma. This chapter delves into the principles, design, and applications of plasma gasifiers, highlighting their benefits, challenges, and the role of plasma modelling in optimizing their performance.

8.1 Introduction to Plasma Gasification

Plasma gasification is a high-temperature process that uses plasma to convert organic and inorganic materials into syngas, which is a mixture of hydrogen, carbon monoxide, and other trace gases. The plasma is generated by electric arcs or microwave energy, creating a highly ionized gas with temperatures exceeding 10,000°C. This intense heat breaks down the feedstock into its basic molecular components.

Applications: Plasma gasifiers are used in waste management to treat municipal solid waste, hazardous waste, and medical waste. They are also applied in energy production by converting biomass into syngas, which can be used to generate electricity or produce synthetic fuels.

8.2 Principles of Plasma Gasification

The plasma gasification process can be divided into several stages:

1. **Feedstock Preparation**: Waste materials are shredded and dried to ensure uniformity and optimal conditions for gasification.
2. **Plasma Generation**: Electric arcs or microwave energy ionize the gas (usually air, oxygen, or steam), creating plasma.
3. **Gasification Reaction**: The feedstock is introduced into the plasma reactor, where it undergoes pyrolysis and gasification. The extreme temperatures break down

theorganic material into syngas and inorganic material into vitrified slag.

4. **Syngas Cleanup**: The syngas produced is cooled and cleaned to remove particulates, tar, and other contaminants before it is used for energy production or chemical synthesis.

Chemical Reactions: The primary reactions involved in plasma gasification include pyrolysis, oxidation, and reduction:

- **Pyrolysis**: $C_xH_y \rightarrow C + H_2 +$ Hydrocarbons
- **Oxidation**: $C + O_2 \rightarrow CO_2$
- **Reduction**: $C + H_2O \rightarrow CO + H_2$

These reactions occur simultaneously within the plasma gasifier, converting the feedstock into syngas and inert slag.

8.3 Design and Components of Plasma Gasifiers

Plasma Torch: The core component of a plasma gasifier is the plasma torch, which generates high-temperature plasma. There are different types of plasma torches, including direct current (DC) arc, alternating current (AC) arc, and radiofrequency (RF) torches.

Reactor Chamber: The reactor chamber is where the gasification reactions take place. It is designed to withstand high temperatures and corrosive environments. The chamber is usually lined with refractory materials to protect it from thermal and chemical wear.

Feedstock Handling System: This system includes mechanisms for shredding, drying, and feeding the waste materials into the reactor. It ensures a continuous and controlled supply of feedstock to the gasifier.

Syngas Cleaning System: After gasification, the syngas passes through a cleaning system to remove impurities. This system typically includes cyclones, scrubbers, and filters to ensure the syngas meets the required purity standards for downstream applications.

Figure 17. An industrial design of a plasma gasifier system

8.4. Modelling and Simulation of Plasma Gasifiers

Importance of Modelling: Plasma gasification involves complex physical and chemical processes, making it challenging to optimize through experimentation alone. Modelling and simulation provide valuable insights into the behavior of the gasifier, allowing for the optimization of design and operational parameters.

Numerical Methods: The modelling of plasma gasifiers typically involves computational fluid dynamics (CFD) and chemical kinetics simulations. These methods help predict temperature distributions, flow patterns, and reaction kinetics within the gasifier.

Example COMSOL Multiphysics Model (MATLAB):

```
geom1 = model.geom.create('geom1', 3);
geom1.lengthUnit('m');
```

```
geom1.create('blk1', 'Block').set('size', [1, 1, 3]);
geom1.create('cyl1', 'Cylinder').set('r', 0.1).set('h', 3);
geom1.run;

model.physics.create('plasma', 'Plasma', 'geom1');
model.physics('plasma').feature.create('term1', 'Terminal', 2).selection.set([1]);
model.physics('plasma').feature.create('gnd1', 'Ground', 2).selection.set([2]);
model.study.create('std1');
model.study('std1').create('time', 'Transient');
model.study('std1').feature('time').set('tlist', 'range(0,0.01,1)');
model.study('std1').feature('time').activate('plasma', true);

model.sol.create('sol1');
model.sol('sol1').study('std1');
model.sol('sol1').create('st1', 'StudyStep').study('std1');
model.sol('sol1').feature('st1').create('fc1', 'FullyCoupled');
model.sol('sol1').feature('st1').create('i1', 'Iterative');
model.sol('sol1').feature('st1').feature('i1').set('linsolver', 'gmres');
model.sol('sol1').runAll;
model = ModelUtil.create('Model');
```

Simulation Results: The results from plasma gasifier simulations can provide insights into:

- **Temperature Distribution**: Understanding how temperature varies within the reactor helps optimize the placement of feedstock and plasma torches.
- **Flow Patterns**: Analysing the flow of gases and particulates can reveal areas where improvements can be made to enhance mixing and reaction efficiency.
- **Reaction Kinetics**: Modelling the chemical reactions allows for optimization of operational parameters, such as the feedstock composition and plasma power, to maximize syngas production and minimize unwanted byproducts.

8.5 Real-World Applications and Success Stories

Waste-to-Energy Plants: Plasma gasifiers are used in waste-to-energy plants to convert municipal solid waste into syngas, which is then used to generate electricity or produce synthetic fuels.

- **Success Story**: The Tees Valley Renewable Energy Facility in the UK utilizes plasma gasification to process up to 950,000 tonnes of waste per year, producing enough electricity to power over 100,000 homes. The facility has demonstrated the feasibility and efficiency of large-scale plasma gasification for waste management [95].

Hazardous Waste Treatment: Plasma gasifiers can safely treat hazardous and medical waste, reducing its volume and toxicity.

- **Success Story**: A plasma gasification plant in Japan processes industrial and medical waste, converting it into syngas and vitrified slag. This plant has successfully reduced the volume of hazardous waste and minimized the release of harmful emissions, showcasing the environmental benefits of plasma gasification [96].

Biomass Gasification: Converting biomass into syngas using plasma gasifiers is an emerging technology for renewable energy production.

- **Success Story**: The Westinghouse Plasma Gasification Plant in Canada uses forestry residues and agricultural waste to produce syngas, which is then converted into biofuels. This plant has demonstrated the potential of plasma gasification to contribute to sustainable energy solutions [97].

8.5.1. Challenges and Opportunities

High Energy Consumption: Plasma gasification requires significant energy input to generate and maintain the plasma, which can impact the overall efficiency and cost-effectiveness of the process.

- **Opportunity**: Advances in plasma torch technology and energy recovery systems can help reduce energy consumption and improve the economic viability of plasma gasifiers.

Feedstock Variability: The heterogeneous nature of waste feedstocks can lead to variations in gasification performance and syngas quality.

- **Opportunity**: Developing robust feedstock preparation systems and adaptive control strategies can help manage feedstock variability and ensure consistent performance.

Economic Viability: The initial capital costs of plasma gasifiers can be high, posing challenges for widespread adoption.

- **Opportunity**: Scaling up production, improving efficiency, and integrating plasma gasifiers with other waste management and energy production systems can enhance their economic attractiveness.

8.5.2. Future Trends in Plasma Gasification

Integration with Renewable Energy Systems: Plasma gasifiers can be integrated with renewable energy systems, such as solar or wind power, to provide a stable and efficient method for converting waste into energy.

- **Advanced Materials and Coatings**: Developing new materials and coatings for reactor components can improve the durability and efficiency of plasma gasifiers, reducing maintenance costs and extending operational lifetimes.
- **Digital Twins and AI**: Using digital twins and AI for real-time monitoring and control of plasma gasifiers can enhance operational efficiency, predict maintenance needs, and optimize performance.

Conclusion

Plasma gasification represents a transformative technology for waste management and energy production, offering a sustainable solution for converting waste into valuable syngas. Through advanced modelling and simulation, the performance of plasma gasifiers can be optimized, addressing challenges such as energy consumption, feedstock variability, and economic viability. As the technology continues to evolve, plasma gasifiers are poised to play a crucial role in achieving sustainable waste management and renewable energy goals.

9: Medical Waste Destruction Simulation

9.1 Introduction

Plasma is used in medical waste disposal because of its ability to produce high temperatures and active chemical reactions that can destroy hazardous materials. To understand the efficiency of this process, it is important to model the energy distribution of particles in the plasma used.

9.1.1. Basic theory

The energy distribution model of particles in plasma can be based on **Maxwell-Boltzmann** (MB) and **Druyvesteyn** (Dn) distributions, depending on the collision conditions of particles in the plasma. In the disposal of medical waste, we assume the plasma is in collisional conditions (many collisions), so the Druyvesteyn distribution is more relevant.

Energy Distribution Function

Particle energy distribution *f(ϵ)* in plasma can be determined using the velocity distribution. For collisional conditions, we use the modified Druyvesteyn distribution:

$$f_{Dn}(\epsilon) = A_{Dn}\epsilon^{1/2}\exp(-C_{Dn}\epsilon^2)$$

Where:

- A_{Dn} is a normalization constant.
- $C_{Dn} = \alpha C^2$, with α is a constant that depends on the collision frequency and C is an exponential constant.

9.1.2.. Mathematical Model

To construct a model of the energy distribution of particles in a plasma, we use the equations of motion and collisional parameters. Suppose $b = n$ is the ratio between the collision frequencies n and external field frequency ω. The modified

Druyvesteyn energy distribution equation becomes:

$$f_{Dn}^{coll}(\epsilon) = A_{Dn}\epsilon^{1/2}\exp(-C_{Dn}(\beta\epsilon)^2)$$

9.1.3. Simulation Steps

In this section, we outline the essential steps required to simulate plasma behavior with respect to energy distributions and collisions. The goal is to provide a comprehensive framework for understanding how key plasma parameters influence energy distributions under the effects of an external field and collisional interactions. The simulation follows a structured process to ensure accurate modeling of particle dynamics.

1. **Define Plasma Parameters**:

   The first step is to define critical plasma parameters. These include:

   - **Electron temperature (*T*)**: Governs the thermal motion of electrons within the plasma.
   - **Collision frequency (*n*)**: Describes the frequency of collisions between particles, which significantly affects energy transfer.
   - **External field frequency (*ω*)**: The frequency of the applied external electromagnetic field, influencing particle motion.
   - **Collisional parameter (*b=n/ω*)**: This parameter represents the ratio of collision frequency to the external field frequency and serves as an important factor in determining the collisional dynamics of the plasma.

2. **Calculate Normalization Constant**:

   The next step involves calculating the normalization constant A_{Dn} , which is determined using the normalization conditions. This ensures that the particle energy distribution adheres to conservation laws and accurately represents the overall population of particles within the plasma.

$$\int_0^\infty f_{Dn}^{coll}(\epsilon)\, d\epsilon = 1$$

3. **Create Energy Distribution Plots**:

Using the Wolfram Language, particle energy distributions are modeled based on the previously defined plasma parameters. This step generates visual plots that illustrate how energy is distributed among the particles, taking into account the effects of collisions and external fields.

Wolfram Language Implementation

```
(* Parameter *)
te = 10000; (* Suhu       in Kelvin *)
nu = 1*10^6; (* Frekuensi tabrakan  dalam Hz *)
omega = 2*10^6; (* Frekuensi medan eksternal dalam Hz *)
beta = nu / omega;
me = 9.10938356*10^-31; (* Massa elektron dalam kg *)
kB = 1.380649*10^-23; (* Konstanta Boltzmann dalam J/K *)

(* Konstanta Eksponensial *)
c = me / (2 * kB * te);
alpha = (Gamma[1/4]^4) / (72 Pi^2);
cdn = alpha * c^2;

(* Fungsi   Distribusi   Energi Druyvesteyn yang  is Dimodifikasi *)
fdnColl[epsilon_] := (1 / NormalizationConstant) * epsilon^(1/2) * Exp[-cdn *
(beta * epsilon)^2]

(* Konstanta Normalisasi dengan Batas Energi Tertentu *)
NormalizationConstant = NIntegrate[epsilon^(1/2) * Exp[-cdn * (beta *
epsilon)^2], {epsilon, 0, 10^-18}, WorkingPrecision -> 25]

(* Plot Distribusi   Energi *)
Plot[fdnColl[epsilon], {epsilon, 0, 5*10^-18},
 PlotRange -> All,
 AxesLabel -> {"Energi (J)", "Distribusi   Energi"},
 PlotLabel -> "Distribusi   Energi Partikel Plasma"]
```

These steps collectively provide a foundation for simulating plasma dynamics, allowing researchers to explore the complex interplay between temperature, collisions, and external forces in shaping particle energy distributions.

9.2 Discussion

This energy distribution model can be used to predict the behaviour of particles in plasmas used for medical waste destruction. By knowing the energy distribution, we can optimize the plasma operating conditions to maximize waste destruction efficiency. Parameters such as electron temperature, collision frequency, and external field frequency greatly affect the energy distribution and must be adjusted accordingly.

Conclusion

Modeling the energy distribution of particles in plasma using the modified Druyvesteyn distribution provides important insights for medical waste disposal applications. This simulation helps in understanding how plasma particles interact with waste materials and can be used to optimize the disposal process.

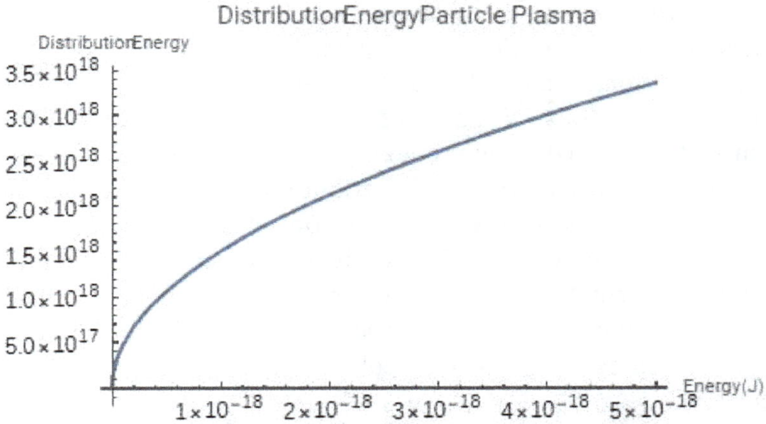

Figure 18. Implementation Output simulation: Energy Distribution of Plasma particle

x

10: Modelling Sheath Plasmas in Comparisson of Maxwell-Boltzmann and Druyvesteyn Distribution Functions

This section reviews distribution function of Sheath Plasma. The research [98] focuses on comparing the Maxwell-Boltzmann (MB) and Druyvesteyn (Dn) distribution functions in the context of plasma sheaths, particularly in partially ionized plasmas. The MB distribution, commonly used to describe the velocity and energy distributions of particles in a non-collisional regime, assumes that the collision frequency is independent of particle velocity. In contrast, the Dn distribution is more suited for collisional regimes where the collision frequency is directly proportional to the velocity. This study extends the understanding of these distributions by analyzing their behavior in both effusional (where particles escape from a region) and non-effusional cases, considering collisional and non-collisional conditions.

The research introduces a collisional parameter β, defined as the ratio of the collisional frequency ν to the frequency ω of the applied external electric field. This parameter plays a crucial role in modifying the Dn distribution function for collisional circumstances. The study demonstrates that in the collisional regime, the Dn distribution function shows significant differences from the MB distribution, particularly in the energy distribution of plasma particles. The Dn distribution yields a higher energy distribution in effusional cases compared to non-Maxwellian effusional distributions, highlighting its importance in accurately modeling the behavior of plasmas under these conditions.

The study also emphasizes the significance of the pre-sheath and sheath regions in plasma processing applications, such as etching and deposition. Understanding the energy distribution of particles in these regions is critical for controlling processes like sputtering and material removal. By comparing the MB and Dn

distributions, the research provides insights into how different collisional behaviors can affect plasma processing outcomes, potentially guiding the optimization of these industrial processes.

In conclusion, this research enhances the theoretical understanding of plasma particle distributions by incorporating collisional effects into the Druyvesteyn distribution function, offering a more accurate description of particle behavior in sheath plasmas. This work sets the stage for experimental verification and further exploration of the Dn distribution in practical applications, contributing to the broader field of plasma physics and its industrial applications.

10.1 Differences Between Druyvesteyn and Maxwell-Boltzmann Distributions in Plasma Applications

The Druyvesteyn and Maxwell-Boltzmann (MB) distribution functions describe the velocity and energy distributions of particles in plasmas but differ fundamentally in their assumptions and applications. The MB distribution is based on the premise that particle velocities are not dependent on the frequency of collisions. It is typically used in non-collisional plasmas, where particles, such as electrons, move freely without frequent interactions with other particles or fields. The MB distribution is characterized by an exponential decay of the form

$$f_{MB} \sim \exp(-Cv^2),$$

where C is a constant related to the particle's mass and temperature.

In contrast, the Druyvesteyn (Dn) distribution is used in collisional plasmas, where the collision frequency is proportional to the particle velocity. This distribution assumes a non-Maxwellian behavior, where the velocity distribution has a different functional form, specifically

$$f_{Dn} \sim \exp(-C'v^4).$$

The D_n distribution accounts for the fact that in collisional plasmas, particles frequently interact with other particles, leading to a velocity distribution that deviates from the Maxwellian. As a result, the D_n distribution often shows a flatter and broader distribution compared to the MB distribution, reflecting the increased likelihood of particles having higher or lower velocities than what the MB distribution would predict.

These differences have practical implications in plasma applications, such as in material processing (etching, deposition), where the accurate modeling of particle behavior is crucial. For instance, in regions of the plasma like the sheath and pre-sheath, where collisions are more frequent, the D_n distribution provides a more accurate description of the energy distribution of particles, which in turn influences the efficiency and outcome of processes like **sputtering** or **ion implantation**.

Python implementation of energy distribution comparisons between Maxwell and Druyvesteyn

```
import numpy as np
import matplotlib.pyplot as plt

# Constants for Maxwell-Boltzmann (MB) and Druyvesteyn (Dn) distributions
k_B = 1.38e-23  # Boltzmann constant in J/K
T = 300  # Temperature in Kelvin
m_e = 9.11e-31  # Electron mass in kg
eV_to_Joule = 1.60218e-19  # Conversion factor from eV to Joules

# Maxwell-Boltzmann distribution function
def maxwell_boltzmann(energy, C):
    return np.sqrt(2/np.pi) * (C * energy)**0.5 * np.exp(-C * energy)

# Druyvesteyn distribution function
def druyvesteyn(energy, C):
    return 2 * (C * energy) * np.exp(-C * energy**2)

# Energy range (in eV)
energy = np.linspace(0, 0.25, 1000)  # Energy from 0 to 10 eV

# Convert energy to Joules for the calculation
energy_Joules = energy * eV_to_Joule

# Constants for the distributions
C_MB = 1/(k_B * T)
C_Dn = 1/((k_B * T)**2)

# Calculate the distributions
MB_distribution = maxwell_boltzmann(energy_Joules, C_MB)
Dn_distribution = druyvesteyn(energy_Joules, C_Dn)
```

```
# Normalize the distributions for comparison
MB_distribution /= np.max(MB_distribution)
Dn_distribution /= np.max(Dn_distribution)

# Plotting the distributions
plt.figure(figsize=(10, 6))
plt.plot(energy, MB_distribution, label='Maxwell-Boltzmann Distribution', color='blue')
plt.plot(energy, Dn_distribution, label='Druyvesteyn Distribution', color='red')
plt.title('Energy Distribution Comparison: Maxwell-Boltzmann vs Druyvesteyn')
plt.xlabel('Energy (eV)')
plt.ylabel('Normalized Probability Density')
plt.legend()
plt.grid(True)
plt.show()
```

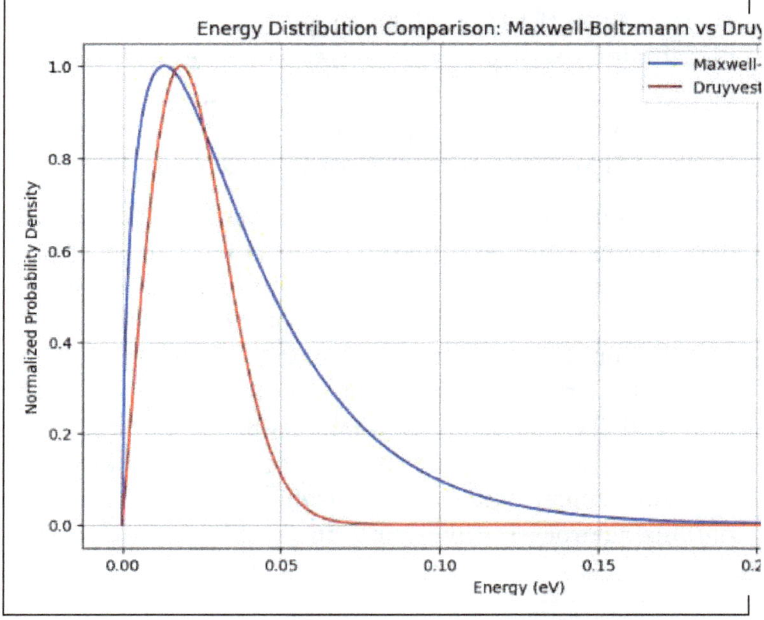

10.2 The Role of Collision Frequency in Differentiating the Maxwell-Boltzmann and Druyvesteyn Distributions

Collision frequency is a critical factor that differentiates the Maxwell-Boltzmann (MB) and Druyvesteyn (D_n) distribution functions in plasma physics. The MB distribution assumes that particles move freely with a velocity distribution that does not depend on the frequency of collisions. This assumption holds true in non-collisional plasmas, where interactions between particles are infrequent, allowing the particles to follow a velocity distribution that depends solely on their thermal energy. In this regime, the MB distribution is characterized by a simple exponential decay in velocity space, leading to a corresponding energy distribution that decreases exponentially with increasing energy.

On the other hand, the Druyvesteyn distribution is applicable in collisional plasmas, where the collision frequency is proportional to the particle velocity. This direct relationship between velocity and collision frequency means that as particles move faster, they are more likely to collide with other particles or neutral atoms. This effect results in a velocity distribution that deviates from the Maxwellian shape, leading to a broader, flatter distribution, particularly at higher velocities. Consequently, the energy distribution in the Dn function declines more rapidly than in the MB distribution, especially at high energies, reflecting the increased likelihood of energy loss through collisions.

In summary, collision frequency plays a fundamental role in determining whether a plasma's particle distribution follows the MB or Dn model. In low-collision environments, the MB distribution prevails, while in high-collision environments, the Dn distribution becomes more accurate, capturing the effects of frequent particle interactions on the velocity and energy distributions. This distinction is essential in applications such as plasma processing, where understanding and predicting particle behavior under different collisional regimes is crucial for optimizing processes like sputtering, etching, and deposition.

```
import numpy as np
import matplotlib.pyplot as plt

# Constants
k_B = 1.38e-23  # Boltzmann constant in J/K
T = 300  # Temperature in Kelvin
m_e = 9.11e-31  # Electron mass in kg

# Maxwell-Boltzmann distribution function (velocity independent of collision
frequency)
def maxwell_boltzmann_velocity(v, C):
    return np.sqrt(2/np.pi) * C**0.5 * v**2 * np.exp(-C * v**2)
```

```
# Druyvesteyn distribution function (velocity proportional to collision frequency)
def druyvesteyn_velocity(v, C):
    return 2 * C * v**2 * np.exp(-C * v**4)

# Velocity range (in m/s)
v = np.linspace(0, 4e5, 1000)  # Velocity from 0 to 1,000,000 m/s

# Constant for the distributions
C_MB = m_e / (2 * k_B * T)
C_Dn = (m_e / (2 * k_B * T))**2

# Calculate the distributions
MB_distribution_velocity = maxwell_boltzmann_velocity(v, C_MB)
Dn_distribution_velocity = druyvesteyn_velocity(v, C_Dn)

# Normalize the distributions for comparison
MB_distribution_velocity /= np.max(MB_distribution_velocity)
Dn_distribution_velocity /= np.max(Dn_distribution_velocity)

# Plotting the distributions
plt.figure(figsize=(10, 6))
plt.plot(v, MB_distribution_velocity, label='Maxwell-Boltzmann Distribution', color='blue')
plt.plot(v, Dn_distribution_velocity, label='Druyvesteyn Distribution', color='red')
plt.title('Velocity Distribution Comparison: Maxwell-Boltzmann vs Druyvesteyn')
plt.xlabel('Velocity (m/s)')
plt.ylabel('Normalized Probability Density')
plt.legend()
plt.grid(True)
plt.show()
```

This script will produce a graph that compares the velocity distributions of particles according to the Maxwell-Boltzmann and Druyvesteyn distributions, clearly showing the impact of collision frequency on the Druyvesteyn distribution. The Maxwell-Boltzmann distribution will appear as a smooth curve, while the Druyvesteyn distribution will show a sharper decline at higher velocities due to the influence of collision frequency.

10.3 Maxwell-Boltzmann and Druyvesteyn Distributions for Plasma Gasification of Municipal Solid Waste (MSW)

Plasma gasification is an advanced thermal treatment process that converts municipal solid waste (MSW) into syngas (a mixture of hydrogen and carbon monoxide) and slag through the application of high-temperature plasma. Understanding the behavior of plasma particles, particularly electrons, in this process is crucial for optimizing the efficiency and output of the gasification process. The Maxwell-Boltzmann (MB) and Druyvesteyn (Dn) distributions provide different perspectives on particle behavior in the plasma environment, especially regarding the role of collision frequency.

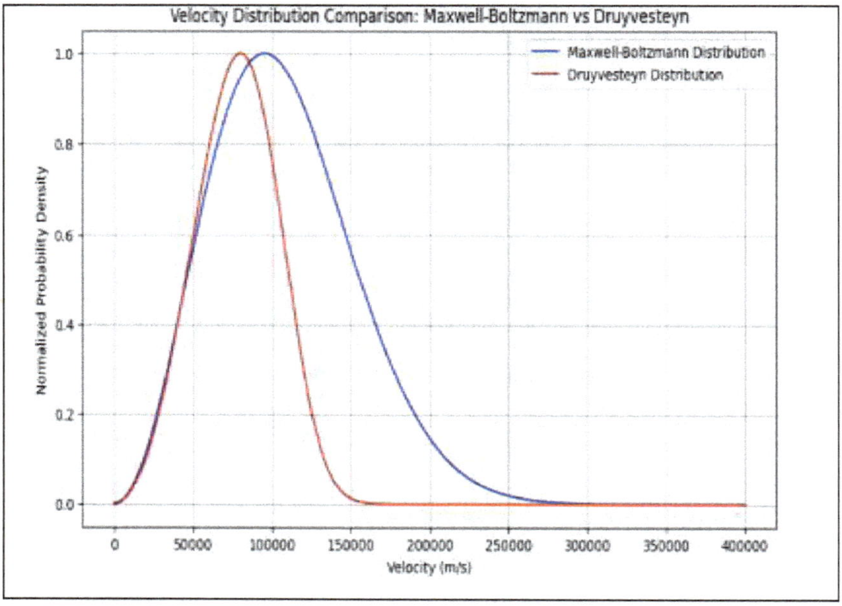

Maxwell-Boltzmann Distribution: In the context of plasma gasification, the MB distribution is applicable in regions of the plasma where collisions are infrequent, such as in high-temperature zones where the mean free path of particles is long. Here, the electron velocity distribution follows a Maxwellian shape, meaning that most electrons have velocities around a certain average value, with fewer electrons having much higher or lower velocities. This distribution is important in understanding the bulk behavior of the plasma, particularly in determining the thermal equilibrium and the temperature of the electrons, which in turn influences the overall efficiency of the gasification process.

- **Druyvesteyn Distribution**: The Dn distribution, on the other hand, becomes more relevant in regions where collisions are frequent, such as in the boundary layers (sheath regions) near the walls of the reactor or in lower temperature zones where the plasma is interacting more with the MSW material. In these regions, the electron velocity distribution deviates from the Maxwellian shape due to the increased frequency of collisions. The Dn distribution typically predicts a flatter, broader velocity profile with a sharper drop-off at high velocities. This indicates that electrons in these regions are losing energy more rapidly due to collisions, which could affect the ionization rates and the stability of the plasma. These factors are critical for maintaining the necessary plasma conditions for efficient gasification.
- **Implications for Plasma Gasification**: The transition between MB and Dn distributions in different parts of the plasma reactor highlights the importance of controlling the plasma conditions to optimize the gasification process. In high-collision regions, where the Dn distribution dominates, maintaining sufficient energy levels despite frequent collisions is crucial for sustaining the ionization and dissociation processes needed to break down complex organic molecules in the MSW into syngas components. Conversely, in low-collision regions, where the MB distribution is more applicable, ensuring uniform temperature distribution is key to maximizing the efficiency of energy transfer within the plasma.

In summary, the MB distribution is useful for understanding the overall thermal properties of the plasma, while the Dn distribution provides insights into the behavior of electrons in more collisional environments, such as near the reactor walls or in cooler parts of the plasma. Both distributions are essential for a comprehensive understanding of plasma behavior in the gasification of municipal solid waste, and they must be considered together to optimize the process for maximum efficiency and output quality.

10.4 Optimizing a plasma gasification process for maximum efficiency and output quality

Optimizing a plasma gasification process for maximum efficiency and output quality involves several steps, including modeling the physical and chemical processes within the gasifier, simulating different operating conditions, and analyzing the results to identify optimal parameters. Here's an outline of the approach, including a basic simulation strategy:

1. Define the System and Key Parameters

- **Feedstock Composition**: Identify the composition of medical waste, including plastics, organics, metals, and other materials.
- **Plasma Characteristics**: Define the properties of the plasma, including temperature, density, and type (e.g., air, argon, etc.).
- **Reactor Design**: Consider the design of the gasifier, including the type of plasma torch, reactor size, and the flow rates of the feedstock and plasma.

2. Model the Gasification Process

- **Energy Balance**: Create an energy balance to model the input energy (from the plasma) and the energy required to gasify the medical waste.
- **Chemical Reactions**: Model the chemical reactions taking place, including the breakdown of hydrocarbons, gasification reactions, and slag formation.
- **Heat Transfer**: Include heat transfer within the reactor, considering both conduction and radiation.
- **Mass Flow**: Model the mass flow of gases, including the production of syngas (CO, H_2) and byproducts like tar and char.

3. Simulate Different Operating Conditions

- **Temperature Variations**: Run simulations at different plasma temperatures to identify the optimal range for maximum syngas production and minimum tar formation.

- **Feedstock Input Rate**: Vary the feedstock input rate to determine the maximum throughput that can be processed without compromising syngas quality.
- **Plasma Gas Flow Rate**: Adjust the flow rate of the plasma gas to optimize the energy transfer to the feedstock.
- **Reactor Pressure**: Simulate different reactor pressures to understand their impact on reaction kinetics and syngas yield.

4. Evaluate and Optimize

- **Syngas Composition**: Analyze the composition of the syngas for different operating conditions, focusing on the CO/H_2 ratio, and minimizing contaminants like nitrogen oxides and sulfur compounds.
- **Energy Efficiency**: Calculate the energy efficiency of the process under different conditions, aiming for the highest ratio of energy output (in the form of syngas) to energy input (from the plasma).
- **Byproduct Minimization**: Evaluate the production of byproducts like slag and tar, optimizing conditions to minimize their formation.

5. Implement Control Strategies

- **Real-time Monitoring**: Develop a control system that monitors key parameters (temperature, pressure, syngas composition) in real-time and adjusts operating conditions to maintain optimal performance.
- **Adaptive Control**: Use adaptive control algorithms that can respond to variations in feedstock composition or other disturbances to maintain consistent output quality.

6. Simulation Tools and Software

- **CFD Software**: Use Computational Fluid Dynamics (CFD) software like ANSYS Fluent or COMSOL Multiphysics to simulate fluid flow, heat transfer, and chemical reactions within the gasifier.
- **Process Simulation Software**: Utilize process simulation tools like Aspen Plus or ChemCAD to model the chemical

processes and perform sensitivity analyses.

- **Optimization Algorithms**: Implement optimization algorithms, such as genetic algorithms or simulated annealing, to find the best combination of operating parameters.

Example Python Simulation for Sensitivity Analysis

Here's a simple Python script for conducting a sensitivity analysis on plasma temperature and feedstock input rate:

```
import numpy as np
import matplotlib.pyplot as plt

# Define ranges for temperature and feedstock input rate
temperatures = np.linspace(2000, 5000, 5)  # Plasma temperatures in Kelvin
input_rates = np.linspace(1, 10, 10)  # Feedstock input rates in kg/h

# Example function to simulate syngas yield (simplified for illustration)
def syngas_yield(temp, rate):
    # Simplified model: syngas yield increases with temperature and input rate,
with diminishing returns
    yield_ = (0.8 * temp / 5000) * (0.9 * rate / 10)
    return yield_

# Perform simulation
yields = np.zeros((len(temperatures), len(input_rates)))

for i, temp in enumerate(temperatures):
    for j, rate in enumerate(input_rates):
        yields[i, j] = syngas_yield(temp, rate)

# Plotting the results
plt.figure(figsize=(10, 6))
for i, temp in enumerate(temperatures):
    plt.plot(input_rates, yields[i, :], label=f'Temperature = {temp} K')

plt.title('Syngas Yield vs. Feedstock Input Rate at Different Temperatures')
plt.xlabel('Feedstock Input Rate (kg/h)')
plt.ylabel('Syngas Yield (arbitrary units)')
plt.legend()
plt.grid(True)
plt.show()
```

7. Interpret the Results

- Use the simulation results to identify the optimal operating temperature and feedstock rate that maximize syngas production and minimize byproducts.

- Validate these findings with experimental data if available, or conduct pilot-scale experiments to confirm the predictions.

8. Continuous Improvement

- As you gather more data from actual operations, refine your model and simulations to improve accuracy.
- Consider implementing machine learning algorithms to predict optimal conditions based on historical data and real-time monitoring.

This approach provides a structured way to optimize the plasma gasification process for medical waste, ensuring high efficiency and high-quality output.

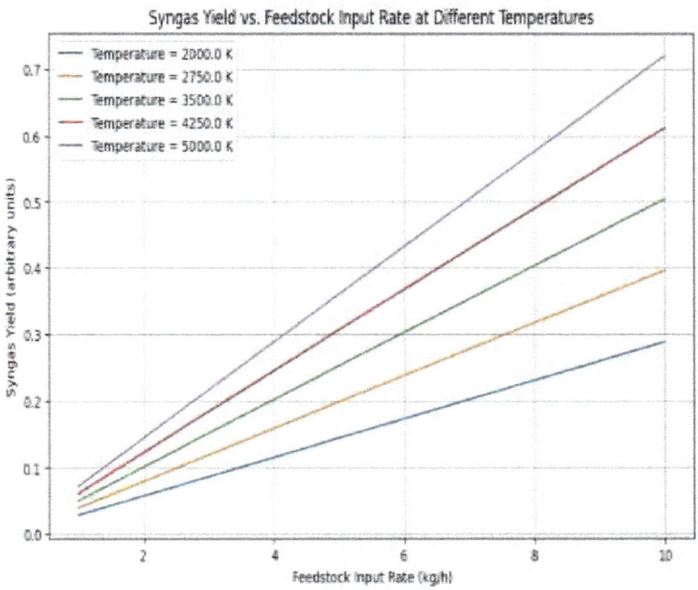

Figure 19. Syngas yield vs feedstock input rate at different temperatures

11: The Municipal Solid Waste Plasma Gasification

The ever-growing accumulation of municipal solid waste (MSW) presents a significant challenge for waste management systems worldwide. As urban populations continue to rise, so does the generation of waste, which not only leads to environmental pollution but also consumes valuable land for waste disposal in landfills. Simultaneously, there is an increasing demand for renewable energy sources to address the global energy crisis and combat climate change. In this context, waste-to-energy (WtE) technologies have gained attention as a promising approach to tackling both the waste management and energy production challenges. Among these technologies, gasification stands out for its potential to convert waste into valuable energy products. This theoretical study explores **plasma gasification** as an advanced and efficient method for converting MSW into syngas, a versatile fuel that can be utilized for energy generation.

11.1. MSW Plasma Gasification

Gasification is a thermal process that converts carbon-based materials, such as waste or biomass, into syngas, a mixture primarily composed of hydrogen (H_2), carbon monoxide (CO), and carbon dioxide (CO_2). The gasification process typically occurs in several stages, including **drying**, **pyrolysis**, **combustion**, and **gasification**.

- **Drying**: MSW is first heated to evaporate moisture, ensuring that the feedstock is suitable for the subsequent thermal processes.
- **Pyrolysis**: At this stage, the dried MSW is subjected to high temperatures in the absence of oxygen, leading to the breakdown of complex organic materials into volatile gases and char.

- **Combustion**: Some of the volatile gases generated during pyrolysis undergo partial oxidation, providing the necessary heat for the gasification reactions.
- **Gasification**: The remaining carbonaceous material reacts with limited amounts of oxygen, steam, or other gasifying agents to produce syngas.

While conventional gasification technologies have been employed for decades, they often face limitations in efficiency, emissions, and the quality of the produced syngas. This is where **plasma gasification** enters the picture as a potential game-changer for waste-to-energy applications.

11.1.1. Plasma Gasification: An Innovative Approach

Plasma gasification employs a plasma torch to generate extremely high temperatures (ranging from 3,000°C to 7,000°C) to break down waste materials at the molecular level. A plasma torch is created by applying an electric current to a gas, which ionizes the gas and generates plasma—an electrically conductive, highly energetic state of matter. This high-temperature plasma environment facilitates the complete decomposition of complex waste materials, even those that are otherwise difficult to process, such as plastics, metals, and hazardous substances.

The advantage of using plasma in the gasification process lies in the high energy density and precision control it offers. The high temperatures achieved in plasma gasification ensure the breakdown of even the most resistant materials into simpler molecules, leading to a more efficient conversion of waste into syngas. Additionally, plasma gasification produces minimal residuals in the form of vitrified slag, a glass-like, non-toxic material that can be safely used in construction or disposed of without environmental harm.

The research conducted by Sato, Abe, and Kato (2010) focuses on the application of plasma gasification for MSW and highlights the process's potential for producing high-quality syngas with fewer impurities compared to traditional gasification methods[97].

11.1.2. Plasma Gasification Process

The plasma gasification process begins with the preparation of the MSW feedstock, which is typically shredded or pre-treated to ensure uniformity. The feedstock is then fed into a gasifier, where it is exposed to the high-energy plasma torch. The gasifier is typically designed to operate in an oxygen-starved environment to promote gasification reactions rather than combustion.

Once inside the gasifier, the high temperatures generated by the plasma torch lead to the rapid breakdown of the MSW into its basic chemical constituents. Organic materials, such as plastics and biomass, decompose into hydrogen, carbon monoxide, and small amounts of methane, while inorganic materials, such as metals and glass, melt and form slag.

The syngas produced during plasma gasification can then be cleaned and processed for various applications. Depending on the composition of the waste and the operating conditions, the syngas may contain varying amounts of hydrogen, carbon monoxide, carbon dioxide, and trace gases. The syngas can be used directly as a fuel for power generation or can be further refined to produce hydrogen, synthetic natural gas, or liquid fuels.

11.1.3. Benefit of Plasma Gasification

Plasma gasification offers several advantages over conventional gasification and other waste-to-energy technologies:

1. **Higher Efficiency**: The extreme temperatures achieved in plasma gasification ensure the complete breakdown of waste materials, resulting in a more efficient conversion of waste to syngas. This also minimizes the formation of

tar and other unwanted byproducts, which are common in lower-temperature gasification processes [97].

2. **Cleaner Syngas**: Plasma gasification produces syngas with fewer contaminants, such as nitrogen oxides (NOx) and sulfur compounds, reducing the need for extensive gas cleaning systems. This makes the syngas more suitable for use in fuel cells or other energy applications that require high-purity fuel.
3. **Reduced Landfill Usage**: By converting waste into energy and slag, plasma gasification significantly reduces the volume of waste that would otherwise be sent to landfills. The slag produced is inert and can be used in construction, further minimizing environmental impact.
4. **Flexibility**: Plasma gasification can process a wide range of waste materials, including hazardous waste, medical waste, and electronic waste, in addition to MSW. This flexibility makes it a versatile technology for waste management.
5. **Lower Emissions**: The controlled environment in a plasma gasifier reduces the formation of harmful emissions, such as dioxins and furans, which are commonly associated with incineration. Plasma gasification also has the potential to reduce greenhouse gas emissions when compared to traditional waste disposal methods.

11.1.4. Risk and Considerations

Despite its numerous advantages, plasma gasification is not without its challenges. One of the main obstacles to the widespread adoption of plasma gasification is the high initial capital cost of constructing plasma gasification facilities. The technology requires significant investment in specialized equipment, such as plasma torches and high-temperature gasifiers, as well as sophisticated gas cleaning systems.

Additionally, plasma gasification is an energy-intensive process, and its overall energy efficiency depends on the design of the system and the integration of energy recovery technologies. For plasma gasification to be economically viable, it is essential to optimize the process to maximize energy recovery from the produced syngas.

Another consideration is the scalability of plasma gasification. While the technology has been demonstrated in pilot projects and small-scale facilities, scaling up to process larger volumes of MSW in urban areas presents logistical and technical challenges. Ensuring a consistent supply of waste feedstock and maintaining operational efficiency at larger scales will be critical to the success of plasma gasification as a mainstream waste-to-energy solution.

11.2. Computational Model Approach

The study uses the Aspen Plus simulation software to model the plasma gasification process. The model is based on the Gibbs free energy minimization method, which calculates the equilibrium composition of syngas. The model simulates the entire gasification process, considering different gasifying agents (air, oxygen, steam) and their impact on the final syngas composition.

11.2.1. Materials and Methods

- **Fuel**: The fuel used in the simulations is representative of Portuguese MSW, with its composition divided into organic elements like paper, plastics, glass, etc.
- **Method Validation**: The model's results were validated against existing literature data on plasma gasification, showing reasonable agreement.

11.2.2. Analysis of Results

The results section is divided into several subsections:

- **Equivalence Ratio (ER) Effect**: The study shows how varying the ER (the ratio of oxygen supplied to the stoichiometric amount required) influences the syngas composition, with lower ER values favoring hydrogen production.
- **Steam/MSW Ratio Effect**: Increasing the steam to MSW ratio enhances hydrogen production due to the steam's high hydrogen content.
- **Gasifier Temperature Effect**: Higher gasifier temperatures generally increase the CO concentration in syngas, with slight decreases in CO_2 and H_2.
- **H_2 Production**: The highest hydrogen yield is obtained with low ER values and high steam flow, particularly in an oxygen-rich environment.
- **CO Production**: CO concentration increases with higher temperatures and moderate ER values, with oxygen-rich environments producing more CO.
- **Lower Heating Value (LHV)**: The study finds that the LHV of syngas is highest when oxygen is used as the gasifying agent, with an optimal ER of 1 and a gasifier temperature of 1500°C.
- **Tar Analysis**: The study addresses the issue of tar formation, a common problem in gasification. The high temperatures in plasma gasification significantly reduce tar content, making the process more efficient.

11.3. Conclusion

Plasma gasification represents a promising advancement in waste-to-energy technology, offering an efficient and environmentally friendly method for converting MSW into valuable syngas. By leveraging the high temperatures and precise control provided by plasma torches, plasma gasification can overcome many of the limitations of traditional gasification processes, producing cleaner syngas with fewer impurities. Although challenges such as high capital costs and energy requirements remain, ongoing research and development efforts are likely to address these issues and pave the way for the broader adoption of plasma gasification. As cities continue to seek sustainable solutions for waste management and energy production, plasma gasification has the potential to play a pivotal role in achieving these goals.

The study concludes that plasma gasification is a promising technology for converting MSW into high-quality syngas. The process benefits from high temperatures, which enhance syngas quality by reducing tar content and improving the LHV. The results indicate that using steam as a gasifying agent yields the highest hydrogen production, while oxygen improves CO content. The study suggests further research and economic analysis to optimize the gasification process. This paper contributes to the field of WtE technologies by providing a detailed simulation study of plasma gasification, highlighting its potential advantages over conventional gasification methods.

12: Plasma Gasification of Medical Waste for Energy Recovery

Plasma gasification of medical waste for energy recovery involves applying the principles and findings from the plasma gasification of municipal solid waste (MSW) discussed in the paper you provided, with specific considerations for the unique characteristics of medical waste.

12.1. Characteristics of Medical Waste

Medical waste is typically composed of a variety of materials, including plastics (e.g., syringes, gloves), paper products, textiles (e.g., gowns, bandages), sharps, and sometimes organic materials (e.g., tissues). The composition of medical waste often contains a higher proportion of plastics compared to municipal solid waste, which can impact the gasification process, especially in terms of syngas composition and energy recovery potential.

12.1.1. Plasma Gasification Process for Medical Waste

Plasma gasification is particularly well-suited for treating hazardous materials, such as medical waste, due to its ability to operate at extremely high temperatures (typically between 2500°C and 4500°C). These high temperatures ensure the complete breakdown of complex molecules, including hazardous compounds, into their elemental forms, significantly reducing the production of harmful byproducts such as dioxins and furans.

12.1.2. Syngas Composition

Hydrogen (H_2) Production:

- Given the high plastic content in medical waste, plasma gasification can produce a syngas rich in hydrogen. As seen in the paper, the use of steam as a gasifying agent enhances hydrogen yield, and this would be applicable in the case of medical waste as well. A high steam-to-waste ratio (S/MW) would be beneficial for maximizing hydrogen production.

Carbon Monoxide (CO) Production:

- Plastics in medical waste would contribute to the production of carbon monoxide when gasified in an oxygen-rich environment. The paper shows that moderate ER values optimize CO production, which is critical for enhancing the lower heating value (LHV) of the syngas.

12.1.3. Energy Recovery

The LHV of the syngas produced from medical waste can be higher than that from typical MSW due to the high calorific value of the plastic content. The study in the paper suggests that using oxygen as a gasifying agent at an optimal ER (around 0.6 to 1.0) and maintaining high gasification temperatures (e.g., 1500°C) can result in syngas with a high LHV, potentially around 13 MJ/Nm^3 or higher, depending on the exact composition of the medical waste.

12.1.4. Environmental Considerations

Tar Production:

- Similar to the findings in the paper, the high temperatures in plasma gasification can minimize tar formation, which is a significant advantage over lower-temperature gasification processes. For medical waste, this is particularly important as tars can contain hazardous compounds.

Emission Control:

- Plasma gasification's ability to destroy pathogens and hazardous organic compounds makes it an ideal choice for medical waste treatment. The resultant inert slag can be safely disposed of or used as a construction material, further contributing to the environmental sustainability of the process.

12.1.5. Practical Implementation

Technology Readiness:

- While plasma gasification technology is relatively mature

for MSW, applying it to medical waste requires consideration of regulatory requirements, particularly regarding the handling and disposal of hazardous medical materials.

Economic Viability:

- The economic feasibility of plasma gasification for medical waste hinges on factors such as the availability of waste, the cost of the plasma technology, and the potential revenue from energy recovery. Given the high energy content of medical waste, the energy recovery potential is significant, which could offset operational costs.

Plasma gasification presents a promising solution for the treatment and energy recovery of medical waste. The process can efficiently convert medical waste into a high-energy syngas, while also minimizing environmental impacts such as tar formation and hazardous emissions. The key to successful implementation lies in optimizing the gasification parameters (e.g., ER, temperature, gasifying agent) specific to the composition of medical waste, as well as ensuring compliance with health and environmental regulations. This analysis is based on extrapolating the principles from the MSW plasma gasification study to medical waste. Further empirical research specific to medical waste would be beneficial to refine these insights.

12.2. Optimization Parameters of the Plasma Gasification

Optimizing plasma gasification parameters (e.g., ER, temperature, gasifying agent) for the treatment of medical waste requires a careful balance of the process conditions to maximize energy recovery, minimize harmful byproducts, and ensure compliance with health and environmental regulations. Below is a structured approach to generating a plasma optimization strategy for medical waste gasification.

12.2..1. Understanding Medical Waste Composition

Medical waste is composed of various materials, each with distinct properties and disposal requirements. Plastics, with high energy content, are primarily hydrocarbons. Paper and textiles, made of cellulosic materials, have lower energy content. Organic matter, though a small fraction, is potentially hazardous and requires complete destruction. Metals and glass are non-combustible and do not contribute to syngas but can affect the quality of slag produced during incineration.

Plastics:

- **High Energy Content**: Plastics in medical waste have high energy content because they are primarily composed of hydrocarbons. This means they can generate a significant amount of energy when incinerated.
- **Composition**: Commonly used in disposable medical items such as syringes, IV bags, gloves, and protective gear.
- **Examples and Chemical Formulas**:
    - Polyethylene (PE): $(C_2H_4)_n$
    - Polypropylene (PP): $(C_3H_6)_n$

Paper and Textiles:

- **Cellulosic Materials**: Paper and textiles are made of cellulosic materials, which have lower energy content compared to plastics. However, they can still contribute to energy production when incinerated.
- **Usage**: Found in packaging, medical clothing, and wrapping materials.
- **Examples and Chemical Formulas**:
    - Cellulose: $(C_6H_{10}O_5)_n$

Organic Matter:

- **Small Fraction**: Organic matter constitutes a small fraction of medical waste but poses significant hazards.
- **Potential Hazards**: Can contain pathogens or other hazardous materials that require complete destruction to prevent infection or contamination.
- **Examples and Chemical Formulas**:
    - Proteins: $C_xH_yO_zN_w$ (general formula)

Metals and Glass:

- **Non-Combustible**: Metals and glass are non-combustible and typically do not contribute to the formation of syngas (synthetic gas) during incineration.
- **Impact on Slag Quality**: Although non-combustible, metals and glass can affect the quality of slag produced during incineration. Slag is the residue left after combustion, and its quality can be influenced by the presence of metals and glass.
- **Examples and Chemical Formulas**:
  - Metals: Iron (Fe), Aluminum (Al)
  - Glass: Silica (SiO_2)

12.2..2. Key Gasification Parameters

Optimizing medical waste gasification involves key parameters such as **Equivalence Ratio** (ER), **temperature**, and **gasifying agents**. A moderate ER (0.4 to 0.6) balances hydrogen and carbon monoxide production, enhancing syngas quality. High temperatures (2000°C to 2500°C) ensure complete breakdown of hazardous materials. Using oxygen and steam as gasifying agents maximizes energy content and syngas quality, making the process efficient and environmentally safe.

a. Equivalence Ratio (ER)

Definition: The ratio of actual oxygen supplied to the amount required for complete combustion.

Optimization:

- Low ER (0.2 to 0.4): Maximizes hydrogen (H2) production but may result in incomplete gasification and higher tar production.
- Moderate ER (0.4 to 0.6): Balances H2 and CO production, improves syngas quality, and reduces tar.
- High ER (0.6 to 1.0): Favours complete gasification with higher CO2 production, reducing syngas energy content

but ensuring thorough destruction of hazardous materials.

- For medical waste, a moderate ER (0.4 to 0.6) is optimal, ensuring good syngas quality with sufficient energy content while minimizing hazardous byproducts.

b. Temperature

- **High Temperature (1500°C to 2500°C)**: Essential for plasma gasification, ensuring complete breakdown of complex molecules, including hazardous substances in medical waste.
- Optimization:
    - **1500°C**: Suitable for producing a balanced syngas composition with a good lower heating value (LHV).
    - **2000°C to 2500°C**: Ensures the complete destruction of hazardous materials, minimizes tar formation, and produces a cleaner syngas.
- For medical waste, temperatures should be maintained at **2000°C to 2500°C** to ensure complete breakdown of hazardous materials and minimize environmental risks.

c. Gasifying Agent

- **Air**: Widely available, but introduces nitrogen, which dilutes the syngas, reducing its energy content.
- **Oxygen (O_2)**: Enhances CO production, resulting in a higher LHV of syngas, but more costly.
- **Steam (H_2O)**: Increases H_2 production, enhancing the energy content of syngas but requires careful control to avoid excessive moisture.

Optimization:

- **Oxygen**: Ideal for high-energy syngas production, particularly when combined with high temperatures.
- **Steam**: Used in combination with oxygen to enhance H2 production, particularly beneficial for medical waste with high plastic content.

A combination of **oxygen and steam** is recommended, with

steam-to-waste ratios (S/MW) optimized to 0.8-1.5 to maximize H2 production while maintaining syngas quality.

12.2..3. Process Monitoring and Control

Effective process monitoring and control are crucial for optimizing medical waste gasification. Implementing real-time monitoring with sensors allows continuous tracking of Equivalence Ratio (ER), temperature, and syngas composition. Automated control systems can then adjust oxygen and steam flow rates based on this real-time data, ensuring optimal gasification conditions. This approach enhances efficiency, safety, and the overall quality of the syngas produced.

- **Real-time Monitoring**: Implement sensors to monitor ER, temperature, and syngas composition continuously.
- **Automated Control Systems**: Use automated systems to adjust oxygen and steam flow rates based on real-time data, ensuring optimal gasification conditions.

12.2..4. Compliance with Health and Environmental Regulations

Ensuring compliance with health and environmental regulations is essential in medical waste gasification. Emission control measures, such as installing scrubbers and filters, and using secondary combustion chambers, help capture harmful compounds. Proper residue management ensures slag is inert and safe for disposal or reuse. Regular environmental and health safety audits maintain adherence to local regulations and standards, promoting a safe and eco-friendly process.

Emission Control:

- Install scrubbers and filters to capture any particulates and volatile organic compounds (VOC_s).
- Utilize a secondary combustion chamber if necessary to ensure the complete destruction of any remaining hazardous compounds.

Residue Management:

- Ensure that the slag produced is inert and free from toxic elements, suitable for safe disposal or reuse.

- **Regular Audits**: Conduct regular environmental and health safety audits to ensure compliance with local regulations and standards.

12.2..5. Economic and Feasibility Considerations

Economic and feasibility considerations are crucial in medical waste gasification. Conduct a cost-benefit analysis to weigh the expenses of oxygen supply and high-temperature operations against the benefits of producing high-quality syngas and reducing environmental impact. Maximize energy recovery by optimizing syngas composition for use in electricity generation or as a feedstock for chemical production, enhancing overall process efficiency.

- **Cost-Benefit Analysis**: Evaluate the cost of oxygen supply and high-temperature operation against the benefits of high-quality syngas and reduced environmental impact.
- **Energy Recovery**: Maximize energy recovery by optimizing the syngas composition to be used in electricity generation or as a feedstock for chemical production.

12.2..6. Summary of Optimal Conditions

These conditions ensure efficient gasification, producing high-quality syngas while minimizing hazardous byproducts and maximizing energy recovery.

- **ER**: 0.4 to 0.6
- **Temperature**: 2000°C to 2500°C
- **Gasifying Agent**: Oxygen combined with steam (S/MW = 0.8-1.5)

12.2..7. Implementation and Testing

This strategy provides a comprehensive approach to optimizing the plasma gasification of medical waste, focusing on maximizing energy recovery while ensuring the process is environmentally sustainable and compliant with health regulations.

- **Pilot Testing**: Before full-scale implementation, conduct pilot tests with the optimized parameters to validate the model predictions and ensure that the process meets energy recovery and environmental safety goals.

12.3. Energy Distribution in a Plasma Reactor for Medical Waste Extermination

The energy distribution in a plasma reactor for medical waste extermination involves understanding how the energy input is utilized in various processes within the reactor. This includes the breakdown of energy consumption in heating, chemical reactions, and the generation of syngas. Below is a detailed analysis:

12.3..1. Overview of Energy Input in Plasma Gasification

Plasma reactors operate by converting electrical energy into thermal energy through a plasma torch. The high temperature generated by the plasma arc (typically between 2000°C and 4500°C) is used to gasify medical waste, resulting in the production of syngas (a mixture of hydrogen, carbon monoxide, and other gases) and slag (an inert residue).

12.3..2. Energy Distribution in Plasma Gasification Process

The energy input into the plasma reactor is distributed among several key processes:

a. Heating of the Plasma Reactor

Thermal Energy: A significant portion of the energy supplied to the plasma torch is used to maintain the high temperatures required for gasification. This includes heating the reactor walls, waste feedstock, and maintaining the plasma state.

Energy Requirement: Depending on the composition of the medical waste, the energy required for heating can vary. For instance, materials like plastics require more energy to break down compared to organic matter.

Approximate Distribution:

Heating: 40% to 60% of the total energy input.

b. Endothermic Chemical Reactions

Gasification Reactions: Energy is consumed in endothermic reactions that break down complex hydrocarbons into simpler molecules, such as H_2, CO, and CO_2. The key reactions include the

Boudouard reaction ($C + CO_2 \rightarrow 2CO$),

the water-gas shift reaction ($CO + H_2O \rightarrow CO_2 + H_2$), and

steam reforming ($CH_4 + H_2O \rightarrow CO + 3H_2$).

Energy Consumption: The energy required for these reactions depends on the specific composition of the waste and the gasifying agents used. Higher plastic content in medical waste will generally increase the energy required for these reactions.

Approximate Distribution:

- **Chemical Reactions**: 20% to 30% of the total energy input.

12.3.3. Syngas Heating

Syngas Energy Content: Some of the energy is transferred to the syngas produced, increasing its temperature. The energy content of the syngas is a critical factor in determining the overall efficiency of the process. The heating value of the syngas is directly influenced by the temperature and the composition of the syngas (proportion of H_2, CO, CH_4, etc.).

Energy Recovery: The energy stored in the syngas can be recovered through combustion in a gas turbine, engine, or used as a chemical feedstock.

Approximate Distribution:

- **Syngas Heating**: 10% to 20% of the total energy input.

12.3.4. Waste Heat

Heat Loss: Some energy is inevitably lost as waste heat, either through the reactor walls, exhaust gases, or cooling systems. Effective insulation and heat recovery systems can minimize these losses.

Energy Efficiency: Improvements in reactor design, such as better insulation and heat recovery units, can reduce energy losses.

Approximate Distribution:

- **Heat Loss**: 10% to 15% of the total energy input.

12.3..5. Energy Efficiency Considerations

Syngas LHV: The lower heating value (LHV) of the syngas is a critical indicator of the efficiency of energy conversion. For medical waste, the LHV can range from 10 to 13 MJ/Nm^3, depending on the feedstock composition and process parameters.

Plasma Torch Efficiency: The efficiency of the plasma torch in converting electrical energy to thermal energy is a key factor. Typically, plasma torches have an efficiency of 70% to 90%.

Waste-to-Energy Ratio: The energy content of the waste (measured in MJ/kg) versus the energy required for gasification determines the net energy output of the process.

12.3..6. Optimizing Energy Distribution

Pre-heating: Pre-heating the medical waste before it enters the reactor can reduce the energy required by the plasma torch.

Heat Recovery: Implementing heat recovery systems to capture waste heat and reuse it within the process (e.g., for pre-heating feedstock or generating steam) can improve overall energy efficiency.

Optimizing Reactor Design: Improving insulation and minimizing heat losses through better reactor design can increase the proportion of energy utilized in productive reactions.

12.3..7. Practical Application

Pilot Testing: Conducting pilot tests with varying medical waste compositions will provide data on the specific energy distribution for different waste streams.

Real-time Monitoring: Implement sensors to monitor energy consumption in real-time, allowing for dynamic adjustments to optimize energy use.

12.3..8. Compliance with Regulations

Emission Controls: Ensure that the energy distribution analysis includes consideration of emission controls, as any additional energy used for scrubbing or filtering emissions will affect the overall efficiency.

Safety Measures: Implement safety protocols to handle the high temperatures and potential hazards associated with plasma gasification of medical waste.

The energy distribution in a plasma reactor for medical waste extermination is a complex interplay of heating, chemical reactions, syngas generation, and waste heat management. By **optimizing the ER, temperature**, and **choice of gasifying agent**, it is possible to enhance the energy efficiency of the process, maximizing the energy recovered from syngas while ensuring safe and environmentally compliant operation.

12.4. Plasma Thermodynamic Equilibrium Simulation

The simulation results for the plasma gasification process have been generated and are displayed in the provided table (Tabel 1). The optimal conditions for maximizing energy recovery in the form of syngas are as follows:

- **Temperature**: 2500 K
- **Equivalence Ratio (ER)**: 1.0
- **Steam to Waste Ratio (S/MW)**: 2.0
- **Maximum Syngas Energy Output**: 15.43 MJ/s

These conditions provide the highest energy yield in syngas, suggesting that operating the plasma reactor at these settings would be most efficient for energy recovery from medical waste.

Figure 20. An ilustration of a complex interplay of heating, chemical reactions, syngas generation, and waste heat management of plasma reactor for medical waste extermination.

Table 2. Model settings of the most efficient for energy recovery from medical waste

ER	T_reactor (K)	S/MW	Energy Output (MJ/Nm^3)
0.2	1000.0	0.5	0.021
0.2	1000.0	0.5	0.021
0.2	1000.0	0.875	0.021
0.2	1000.0	1.0	0.021
0.2	1000.0	1.25	0.021
0.2	1000.0	1.5	0.021
0.4	1500.0	0.5	0.026
0.4	1500.0	1.0	0.026

ER	T_reactor (K)	S/MW	Energy Output (MJ/Nm^3)
0.4	1500.0	1.5	0.026
0.2	1000.0	1.625	0.021
1.0	2500.0	0.5	0.037
1.0	2500.0	0.875	0.037
1.0	2500.0	1.25	0.037
0.2	1000.0	1.625	0.021
0.2	1000.0	2.0	0.021

Sumber: Penulis

DAFTAR PUSTAKA

[1] Cerri G, Moglie F, Montesi R, Russo P, Vecchioni E. FDTD Solution of the Maxwell–Boltzmann System for Electromagnetic Wave Propagation in a Plasma. IEEE Trans Antennas Propag 2008;56:2584–8. https://doi.org/10.1109/TAP.2008.927505.

[2] Bang J, Chung C. Experimental investigation of the Boltzmann relation for a bi-Maxwellian distribution in inductively coupled plasmas. Phys Plasmas 2009;16:93502. https://doi.org/10.1063/1.3174432.

[3] Carman RJ, Mildren RP. Electron energy distribution functions for modelling the plasma kinetics in dielectric barrier discharges. J Phys D Appl Phys 2000;33.

[4] Amemiya H. Plasma Dispersion for Druyvesteyn Type Velocity Distribution. J Physical Soc Japan 2012;81:74501. https://doi.org/10.1143/JPSJ.81.074501.

[5] Li J, Yuan C, Zhou Z, Kudryavtsev A, Wang X, Chen X. Effects of Druyvesteyn Distribution to Transmission Coefficients in Plasma. 2018 12th International Symposium on Antennas, Propagation and EM Theory (ISAPE), 2018, p. 1–3. https://doi.org/10.1109/ISAPE.2018.8634358.

[6] Davis S, Avaria G, Bora B, Jain J, Moreno J, Pavez C, et al. Kappa distribution from particle correlations in nonequilibrium, steady-state plasmas. Phys Rev E 2023.

[7] Tautz R. A new distribution function for relativistic counterstreaming plasmas. Astrophys Space Sci 2010;330:69–72. https://doi.org/10.1007/s10509-010-0351-9.

[8] Jackson JD. Classical electrodynamics. 3rd ed. New York: Wiley; 1998.

[9] Griffiths DJ. Introduction to electrodynamics. 4th ed. Boston: Pearson; 2013.

[10] Cheng DK. Field and wave electromagnetics. 2nd ed. Boston: Addison-Wesley; 1989.

[11] Landau LD, Lifshitz EM. Electrodynamics of continuous media. 2nd ed. Oxford: Pergamon Press; 1984.

[12] Chen FF. Introduction to plasma physics and controlled fusion. 3rd ed. New York: Springer; 2016.

[13] Nicholson DW. Introduction to plasma theory. New York: John Wiley & Sons; 1983.

[14] Bittencourt JA. Fundamentals of plasma physics. 3rd ed. New York: Springer; 2004.
[15] Braginskii SI. Transport processes in a plasma. In: Leontovich MA, editor. Reviews of plasma physics, New York: Consultants Bureau; 1965, p. 205–311.
[16] Shkarofsky IP, Johnston TE, Bachynski MP. The particle kinetics of plasmas. Reading: Addison-Wesley; 1966.
[17] Krall NA, Trivelpiece AW. Principles of plasma physics. San Francisco: San Francisco Press; 1973.
[18] Freidberg JP. Ideal magnetohydrodynamics. New York: Plenum Press; 1987.
[19] Goedbloed JP, Poedts S. Principles of magnetohydrodynamics. Cambridge: Cambridge University Press; 2004.
[20] Boyd TJM, Sanderson JJ. The physics of plasmas. Cambridge: Cambridge University Press; 2003.
[21] Birdsall CK, Langdon AB. Plasma physics via computer simulation. New York: McGraw-Hill; 1985.
[22] Hockney RW, Eastwood JW. Computer simulation using particles. Bristol: Institute of Physics Publishing; 1988.
[23] Stacey WM. Fusion plasma physics. 2nd ed. Weinheim: Wiley-VCH; 2010.
[24] Kivelson MG, Russell CT. Introduction to space physics. Cambridge: Cambridge University Press; 1995.
[25] Lieberman MA, Lichtenberg AJ. Principles of plasma discharges and materials processing. 2nd ed. Hoboken: Wiley; 2005.
[26] Fridman A, Kennedy LA. Plasma physics and engineering. 2nd ed. Boca Raton: CRC Press; 2011.
[27] Fridman A, Friedman G. Plasma medicine. Hoboken: Wiley; 2013.
[28] LeVeque RJ. Finite difference methods for ordinary and partial differential equations: steady-state and time-dependent problems. Philadelphia: SIAM; 2007.
[29] Press WH, Teukolsky SA, Vetterling WT, Flannery BP. Numerical recipes: the art of scientific computing. 3rd ed. New York: Cambridge University Press; 2007.
[30] Hughes TJR. The finite element method: linear static and dynamic finite element analysis. Mineola: Dover Publications; 2000.
[31] Zienkiewicz OC, Taylor RL. The finite element method. 5th ed. Oxford: Butterworth-Heinemann; 2000.

[32] Versteeg HK, Malalasekera W. An introduction to computational fluid dynamics: the finite volume method. 2nd ed. Harlow: Pearson Education; 2007.
[33] Blazek J. Computational fluid dynamics: principles and applications. 3rd ed. Oxford: Butterworth-Heinemann; 2015.
[34] Smith KR, Mehta S. The Burden of Disease from Indoor Air Pollution in Developing Countries : Comparison of Estimates Prepared by Table of Contents 2000.
[35] Sauer T. Numerical analysis. 2nd ed. Boston: Pearson; 2011.
[36] Saad Y. Iterative methods for sparse linear systems. 2nd ed. Philadelphia: SIAM; 2003.
[37] Trefethen LN, Bau D. Numerical linear algebra. Philadelphia: SIAM; 1997.
[38] Kelley CT. Iterative methods for linear and nonlinear equations. Philadelphia: SIAM; 1995.
[39] Dennis JE, Schnabel RB. Numerical methods for unconstrained optimization and nonlinear equations. Philadelphia: SIAM; 1996.
[40] Hairer E, Wanner G. Solving ordinary differential equations II: stiff and differential-algebraic problems. 2nd ed. Berlin: Springer; 2010.
[41] Ascher UM, Petzold LR. Computer methods for ordinary differential equations and differential-algebraic equations. Philadelphia: SIAM; 1998.
[42] Courant R, Friedrichs K, Lewy H. On the partial difference equations of mathematical physics. IBM J Res Dev 1967;11:215–34.
[43] Strikwerda JC. Finite difference schemes and partial differential equations. 2nd ed. Philadelphia: SIAM; 2004.
[44] Lax PD, Richtmyer RD. Survey of the stability of linear finite difference equations. Commun Pure Appl Math 1956;9:267–93.
[45] Berger MJ, Colella P. Local adaptive mesh refinement for shock hydrodynamics. J Comput Phys 1989;82:64–84.
[46] AB C. COMSOL Multiphysics 2024.
[47] Inc. A. ANSYS Fluent 2024.
[48] Foundation O. OpenFOAM 2024.
[49] Pacheco PS. An introduction to parallel programming. Burlington: Morgan Kaufmann; 2011.

[50] Schroeder W, Martin K, Lorensen B. The visualization toolkit: an object-oriented approach to 3D graphics. 4th ed. Clifton Park: Kitware; 2006.
[51] Lieberman MA, Lichtenberg AJ. Principles of plasma discharges and materials processing. 2nd ed. Hoboken: Wiley; 2005.
[52] Boyd TJM, Sanderson JJ. The physics of plasmas. Cambridge: Cambridge University Press; 2003.
[53] Hirt CW, Nichols BD. Volume of fluid (VOF) method for the dynamics of free boundaries. J Comput Phys 1981;39:201–25.
[54] Roache PJ. Verification and validation in computational science and engineering. Albuquerque: Hermosa Publishers; 1998.
[55] Ferziger JH, Peric M. Computational methods for fluid dynamics. 3rd ed. Berlin: Springer; 2002.
[56] Freidberg JP. Plasma physics and fusion energy. Cambridge: Cambridge University Press; 2008.
[57] Starikovskiy A, Aleksandrov N. Plasma-assisted ignition and combustion. Prog Energy Combust Sci 2013;39:61–110.
[58] AB C. COMSOL Multiphysics 2024.
[59] Inc. A. ANSYS Fluent 2024.
[60] Foundation O. OpenFOAM 2024.
[61] Welch DR, Rose DR, Clark RD, Genoni TJ, Hughes TJ. Implementation of a generalized algorithm for electromagnetic PIC plasma simulation. Comput Phys Commun 2004;164:183–8.
[62] Vay JL, Grote DP, Friedman A, Cohen RH. Novel methods in the Particle-In-Cell accelerator Code-Framework Warp. Comput Sci Discov 2011;4:15002.
[63] Ahrens J, Geveci B, Law C. ParaView: an end-user tool for large data visualization. In: Hansen CD, Johnson CR, editors. The Visualization Handbook, Burlington: Elsevier; 2005, p. 717–32.
[64] Inc. T. Tecplot 360 2024.
[65] MathWorks. MATLAB 2024.
[66] Foundation PS. Python 2024.
[67] Verboncoeur JP. Particle-in-cell simulations of plasmas: Review and advances. Plasma Phys Control Fusion 2005;47:A231. https://doi.org/10.1088/0741-3335/47/5A/015.

[68] Schneider R, Runov A. Challenges in plasma edge fluid modelling. Plasma Phys Control Fusion 2007;49:B111. https://doi.org/10.1088/0741-3335/49/12B/S13.
[69] Stoltz PH, others. Modeling tools for plasmas in the strongly-coupled state. AIP Conf Proc 2012;1421:52–61. https://doi.org/10.1063/1.3686894.
[70] Mishchenko A, others. Tools for burning plasmas: Numerical modeling and simulation. Plasma Phys Control Fusion 2023;65:54001. https://doi.org/10.1088/1361-6587/acb374.
[71] Bonzanini A, others. Foundations of machine learning for low-temperature plasmas. Plasma Sources Sci Technol 2023;32:43002. https://doi.org/10.1088/1361-6595/acd077.
[72] Ekezie F-GC, others. A review of recent advances in cold plasma technology for the food industry: Current applications and future trends. Trends Food Sci Technol 2017;69:46–58. https://doi.org/10.1016/j.tifs.2017.08.007.
[73] Vay J-L, others. WarpX: A new exascale computing platform for beam-plasma simulations. J Comput Phys 2017;348:80–104. https://doi.org/10.1016/j.jcp.2017.07.026.
[74] Babaeva NY, Naidis G V. Modeling of plasmas for biomedicine. Plasma Sources Sci Technol 2017;26:43001. https://doi.org/10.1088/1361-6595/aa681a.
[75] Kutz JN. Deep learning in fluid dynamics. J Fluid Mech 2017;814:1–4.
[76] Shukla PK, Eliasson B. Colloquium: Nonlinear collective interactions in quantum plasmas with degenerate electron fluids. Rev Mod Phys 2011;83:885–906.
[77] Bird RP. Multiscale modeling of plasma processes. J Phys D Appl Phys 2007;40.
[78] Birdsall CK, Langdon AB. Plasma Physics via Computer Simulation. New York: McGraw-Hill; 1985.
[79] Cappello S, Escande DF. Understanding and interpreting 3D effects in fusion plasma simulations. Comput Phys Commun 2013;184:271–5.
[80] Kwon J, Chang C, Choi J. Turbulence in magnetically confined fusion plasmas. Plasma Phys Control Fusion 2018;60:94005.

[81] Connor JW, Hastie RJ, Wilson HR. Magnetohydrodynamic instabilities in tokamak plasmas. Plasma Phys Control Fusion 2004;46.
[82] Hsu S, O'Neil T, Pewsey C. Multiscale modeling of plasma dynamics. J Comput Phys 2016;326:201–18.
[83] Kutz JN. Deep learning in fluid dynamics. J Fluid Mech 2017;814:1–4.
[84] Fridman A, Friedman G. Plasma medicine. Hoboken: Wiley; 2013.
[85] Bae J, Kou X, Liu Y. Uncertainty quantification in plasma simulations using machine learning. Phys Plasmas 2020;27:102303.
[86] Shukla PK, Eliasson B. Colloquium: Nonlinear collective interactions in quantum plasmas with degenerate electron fluids. Rev Mod Phys 2011;83:885–906.
[87] Manfredi G. How to model quantum plasmas. Fields Institute Communications, vol. 46, American Mathematical Society; 2005, p. 263–87.
[88] Markowich PA, Ringhofer C, Schmeiser C. Semiconductor equations. New York: Springer; 1990.
[89] Ono K, Shibahara M, Goto T. Plasma etching of silicon and silicon dioxide in CF4-Ar and SF6-Ar inductively coupled plasmas. Journal of Vacuum Science & Technology A: Vacuum, Surfaces, and Films 1996;14:3038–43.
[90] Frisbee RH. Advanced propulsion for NASA exploration missions: enabling propulsion technologies for affordable, safe, and fast exploration of the solar system. J Propuls Power 2003;19:1129–47.
[91] Freidberg JP. Plasma physics and fusion energy. Cambridge: Cambridge University Press; 2008.
[92] Fridman A, Friedman G. Plasma medicine. Hoboken: John Wiley & Sons; 2013.
[93] Fridman A, Chirokov A, Gutsol A. Non-thermal atmospheric pressure discharges. J Phys D Appl Phys 2005;38.
[94] Pfender E. Thermal plasma processing in the nineties. Pure and Applied Chemistry 1988;60:591–606. https://doi.org/10.1351/pac198860050591.
[95] Tees Valley. Tees Valley Renewable Energy Facility n.d.
[96] Sato K, Abe Y, Kato T. Plasma gasification of medical waste in Japan. J Hazard Mater 2010;181:769–75. https://doi.org/10.1016/j.jhazmat.2010.05.081.

[97] Westinghouse Plasma Corporation. Westinghouse Plasma Corporation n.d.

[98] Tiwari PK, Kumar R, Halder K, Lee YS. Maxwell–Boltzmann and Druyvesteyn Distribution Functions Expressing the Particle Velocity and the Energy in Sheath Plasmas. Journal of Russian Laser Research 2023;44:504–12. https://doi.org/10.1007/s10946-023-10157-3.

GLOSARIUM

- **Plasma**: Often referred to as the fourth state of matter, plasma consists of a collection of free-moving ions and electrons that exhibit collective behavior and are highly responsive to electromagnetic fields.
- **Thermal Plasma**: A type of plasma where the temperatures of electrons, ions, and neutral particles are in equilibrium. Common in stars and artificial processes like plasma arcs.
- **Non-Thermal Plasma**: Plasma where the electron temperature is much higher than that of ions and neutral particles. Used in applications such as fluorescent lamps and air purifiers.
- **Magnetohydrodynamics (MHD)**: A macroscopic model treating plasma as a conducting fluid that interacts with magnetic and electric fields, used to describe the large-scale behavior of plasmas.
- **Particle-in-Cell (PIC) Method**: A computational technique used to simulate plasma by tracking individual particles, which helps study kinetic phenomena like plasma waves and particle interactions.
- **Maxwell's Equations**: Fundamental equations governing how electric and magnetic fields propagate and interact with charges in plasma.
- **Debye Length**: The distance over which electric potentials are shielded in plasma. It characterizes the plasma's ability to screen electric fields.
- **Plasma Frequency**: The natural oscillation frequency of electrons in plasma, influencing wave propagation within the plasma.
- **Kinetic Theory**: A detailed description of plasma, considering the velocity distribution of individual particles, governed by equations like the Boltzmann equation.

- **Two-Fluid Model**: A model that treats electrons and ions in plasma as separate fluids, allowing detailed simulation of their individual dynamics.
- **Numerical Methods**: Techniques such as finite difference, finite element, and finite volume methods used to discretize and solve equations governing plasma behavior.
- **Finite Element Method (FEM)**: A numerical technique dividing a domain into small elements, commonly used for complex geometries in plasma simulations.
- **Explicit and Implicit Solvers**: Numerical methods for solving time-dependent plasma equations. Explicit solvers are simple but require small time steps for stability, while implicit solvers allow larger time steps but are computationally intensive.
- **Adaptive Mesh Refinement (AMR)**: A technique in numerical simulations that dynamically adjusts the resolution of the mesh based on the plasma's behavior to optimize computational resources.
- **Sheath**: The boundary layer formed when plasma interacts with a solid object, like a spacecraft, where the plasma's electric fields and particle densities change.
- **Machine Learning in Plasma Modeling**: The integration of machine learning techniques to improve plasma simulation accuracy and efficiency by optimizing models and interpreting large datasets.

INDEX

AUTHOR BIOGRAPHY

Sugeng Rianto: A lecturer specializing in instrumentation at Brawijaya University, Sugeng Rianto has contributed significantly to teaching and research in environmental physics, medical imaging, modeling and simulation, extended reality (XR), and computational science. His academic focus and research work have had a notable impact on instrumentation technology and its applications.

Muhammad Nurhuda: A lecturer at Brawijaya University since 1990, with a Ph.D. from Universit'at Bielefeld, Germany, Muhammad Nurhuda has contributed to teaching subjects like physics, computational physics, and energy research, particularly in non-conventional energy sources. He is known for his broad teaching scope and dedication to energy research.

Abdurrouf: As an academic and researcher in computational physics and fluid dynamics, Abdurrouf leads the Basic Physics Laboratory at Brawijaya University. He has contributed to both teaching and research, particularly in fields like material and fluid mechanics computation, electrodynamics, and statistical physics. His research on Boltzmann temperature in molecular alignment is notable.

Didik Yudianto: A lecturer specializing in Geophysics Engineering, GIS, and Remote Sensing, Didik Yudianto has played a key role in the academic and entrepreneurial development at Brawijaya University. He contributes by supervising student community service projects and leading the university's Entrepreneurship Unit.

Agus Dardiri: A dedicated physics lecturer, Agus Dardiri has contributed to education through his teaching, research, and community service. He has been recognized with various national awards and is known for his efforts to advance education in Indonesia, believing in education as the foundation for a better future.

Plasma Modelling: Innovations and Applications

Plasma Modelling: Innovations and Applications offers an advanced exploration into the evolving field of plasma science, integrating both theoretical and practical insights. The book bridges the gap between plasma physics fundamentals and cutting-edge computational tools, guiding readers through complex numerical methods and their application in real-world scenarios. By addressing critical topics such as plasma etching, fusion energy, and environmental applications, it highlights the importance of plasma modelling in modern technology, from semiconductor manufacturing to medical waste treatment.

Through detailed case studies and emerging trends, such as machine learning, quantum plasmas, and high-performance computing, the book presents a forward-looking perspective on the future of plasma research. It encourages innovation by demonstrating how plasma behaviour can be simulated to solve intricate problems in energy, space, and health industries. With a strong focus on computational tools like COMSOL, OpenFOAM, and Python, readers are equipped with the practical skills to develop and implement their own simulations.

Ideal for researchers, engineers, and advanced students, this book blends deep theoretical knowledge with real-world applications, inspiring the next generation to push the boundaries of plasma technology and explore new possibilities in science and engineering.

www.ingramcontent.com/pod-product-compliance
Lightning Source LLC
Chambersburg PA
CBHW050524240926
51846CB00003B/689
* 9 7 8 6 2 3 9 7 9 1 7 5 9 *